Design of Francis Turbine and OpenFOAM Methods

Zhiwei Guo · Sailesh Chitrakar · Zhongdong Qian ·
Bhola Thapa

Design of Francis Turbine and OpenFOAM Methods

Springer

Zhiwei Guo
School of Water Resources
and Hydropower Engineering
Wuhan University
Wuhan, China

Zhongdong Qian
School of Water Resources
and Hydropower Engineering
Wuhan University
Wuhan, China

Sailesh Chitrakar
Kathmandu University
Dhulikhel, Nepal

Bhola Thapa
Kathmandu University
Dhulikhel, Nepal

ISBN 978-981-99-8380-3 ISBN 978-981-99-8381-0 (eBook)
https://doi.org/10.1007/978-981-99-8381-0

This Springer imprint is published by the registered company Springer Nature Singapore Pte Ltd.
The registered company address is: 152 Beach Road, #21-01/04 Gateway East, Singapore 189721, Singapore

Paper in this product is recyclable.

Preface

Hydropower is considered as one of the cleanest forms of energy. On an average life-cycle basis, carbon dioxide emission from hydropower is 24 g per kWh, which is much less compared to coal (820 g) or natural gas (490 g). Hydropower is one of the growing industries around the world, especially in South and Central Asia, with more than 350 GW of remaining hydropower potential in this region. China was reported as the first-ranked country to have installed new hydropower capacity in 2021, with the total of 20,842 MW by the Hydropower Status Report (2022). Similarly, Nepal was reported as the fourth-ranked country, with the total new installation of 684 MW, after Canada (924 MW) and India (803 MW). This also shows that hydropower is growing rapidly in this part of world.

Operation and maintenance issues of the turbines of hydropower plants are an inevitable problem in Nepal and China, which is primarily posed by excessive sediment content in rivers. This is usually referred to as a unique challenge of Himalayan Rivers, which are also found in the hydropower plants of South America and some parts of Europe. The sediment concentration in rivers during the monsoon season can be higher than 10000 ppm (1 ppm is equal to 1 mg/liter), with Quartz (a hard mineral found in sand) content more than 50%. De-silting basins available in power plants screen more than 80% of sediments contained in the flow, but small particles (around 0.1 mm) remain suspended and are carried into the power house. When the turbines in the plants during monsoon season operate at maximum capacity, maximum energy can be extracted. However, turbines also strike with the hard sediment particles, causing wear, more technically termed as sediment erosion. Yearly, several million dollars are spent for repair and maintenance of the turbines. Apart from these expenses, the loss of production during overhauling and due to efficiency loss after erosion are indirect costs that are usually not accounted for.

Therefore, this book is focused on providing inputs to the academicians for conducting design and optimization of the hydraulic turbines, so that the turbines can be manufactured to withstand the local sedimentation problems of the power plants. In this part, the focus is mainly on Francis turbines, which contains main dimensions, design and the runners, guide vanes and distributors. Computational fluid dynamics

(CFD) studies are performed on these components with and without including the effect of sediments.

There is a lot of literature devoted to CFD methods for fluid mechanics including rotating situation. It is impossible to cover all in a single book. Therefore, we have presented only the method that are suitable and valuable in relative fields.

The book is jointly written by Wuhan University and Kathmandu University, under the co-operation of Hydropower Research Center for Himalayan Region (HRCHR), which was launched in 2020. The main aim of this co-operation is to contribute to solve the key issues in the development of hydropower resources, to develop core technologies in the development of hydropower, and to develop into an open platform for scientific research co-operation and practical application in the field of water conservancy and hydropower, promote collaborative innovation, and better serve the "Belt and Road" initiatives.

We thank all our present and former students, colleagues, and friends in either Wuhan University or Kathmandu University who helped us to finish this work. We couldn't list all names, but the support from Ole Gunnar Dahlhaug, Hari Prasad Neopane, Qinqin Zhong, and Zilong Zhao is mainly appreciated.

Wuhan, China — Zhiwei Guo
Dhulikhel, Nepal — Sailesh Chitrakar
Wuhan, China — Zhongdong Qian
Dhulikhel, Nepal — Bhola Thapa

Contents

Chapter 1
Introduction

This chapter intends to introduce terminologies and symbols that will be used to design Francis turbine in later chapters. Some basic concepts of the hydropower turbines, their history, and recent studies have been discussed.

1.1 Terminologies and Symbols

1.1.1 Velocities

C = Absolute velocity (it represents the velocity of the flow)

U = Peripheral velocity (also referred as tangential velocity. It represents the velocity of the body. In the case of Francis turbines, it represents the velocity of the runner blades, and in the case of Pelton turbines, it represents the velocity of the buckets. If ω is the angular velocity in rad/sec and r is the radius, $U = \omega.r$)

W = Relative velocity (it represents the velocity of the flow relative to the body in motion. In other words, it is the velocity that is experienced by the moving body. In the case of turbines, the rotating blade's angle is shaped according to this velocity).

1.1.2 Velocity Triangle

A triangle is formed due to the velocity vectors C, U, and W acting in different directions.

Following the vector rule,

Z. Guo et al., *Design of Francis Turbine and OpenFOAM Methods*,
https://doi.org/10.1007/978-981-99-8381-0_1

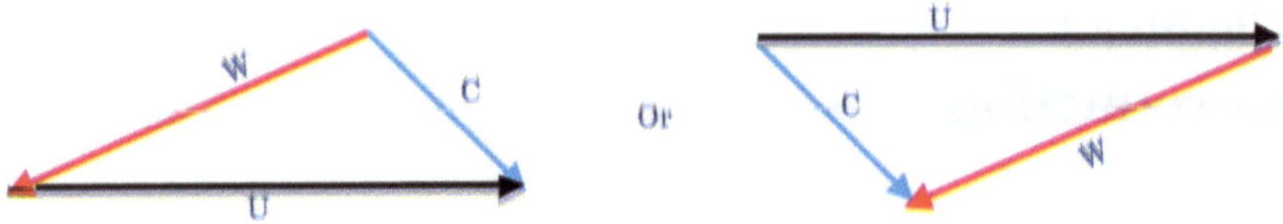

Fig. 1.1 Velocity triangles with different shapes

Fig. 1.2 Velocity triangle with the respective angles

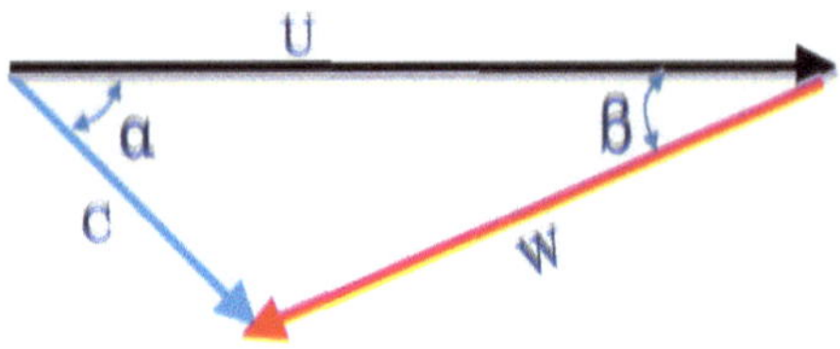

$$\vec{C} = \vec{W} + \vec{U} \tag{1.1}$$

The angles inside the triangle are named as:

α = Angle between U and C

β = Angle between U and W (Figs. 1.1 and 1.2).

1.1.3 Components of Absolute and Relative Velocity

During the 2D design of blades, co-ordinate system *m-u* is taken into consideration. "*u*" represents the tangential direction, and "*m*" represents the meridional direction. In the case of axial flow turbines, "*m*" may be replaced with "*x*", which means the direction of the flow. However in the case of mixed-flow turbines, the direction of flow changes within the runner. Hence, a general representation, "*m*" is used which can be used for all types of turbines.

The absolute velocity can then be divided into its tangential and meridional components.

$C_u = C.\cos\alpha = u$ component of C

$C_m = C.\sin\alpha = m$ component of C.

Similarly, relative velocity can be divided into its respective components.

$W_u = W.\cos\beta$

$W_m = W.\sin\beta$.

In the case of turbines, "*u*" component of the velocity is responsible for conducting the useful work, whereas "*m*" component is used for directing the flow. This is because the rotation of the blades or buckets is in the direction of "*u*" (Fig. 1.3).

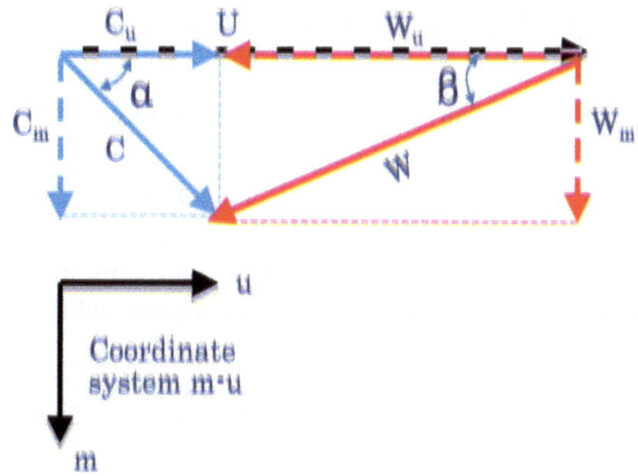

Fig. 1.3 Velocity triangle with respective angles and components in *m-u* plane

1.1.4 Velocity Triangles in a Turbomachine

Any machine that can change the swirl component of the flow, i.e., C_u component while passing from inlet to outlet can be considered as a turbomachine. A small illustration of a turbomachinery blade profile is provided in Fig. 1.4.

In the above figure, the subscripts 1 and 2 are representing the locations at inlet and outlet of any turbomachinery blade respectively. It can be seen from the figure that the tangential component of the flow, i.e., C_u changes as the flow passes into the blade. As the work done by or on the blade depends on the deviation of this component, the above profile represents a turbomachine. It is to be noted that the peripheral velocity U can also be different at inlet and outlet, depending upon their positions relative to the center of the runner. However, the angular velocity ω is same for a given runner at a given time.

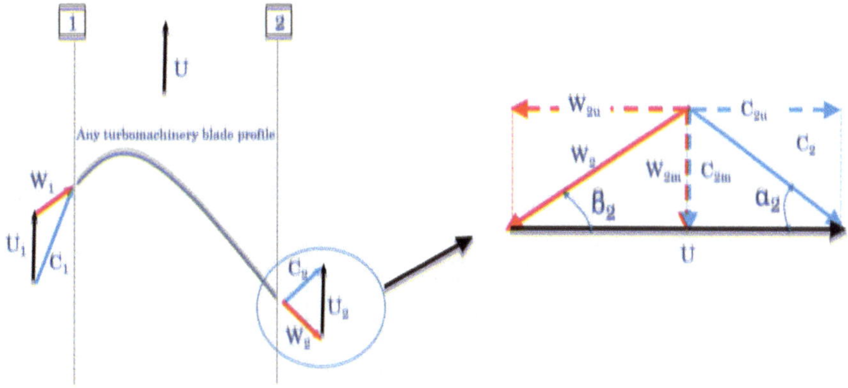

Fig. 1.4 Turbomachinery blade profile indicating the inlet and outlet velocity triangles

1.1.5 Net Head for a Hydraulic Turbine

Referring to Fig. 1.5, the net head for a hydraulic turbine can be defined as [1]:

$$H = \mathrm{EGL_{in}} - \mathrm{EGL_{out}}$$

In the above equation, EGL represents energy grade line. This term can be explained through the Bernoulli equation, which is given by the following relation:

$$\frac{P}{\rho g} + \frac{C^2}{2g} + z = \text{constant} \tag{1.2}$$

where

$\frac{P}{\rho g}$ is the pressure head

$\frac{C^2}{2g}$ is the velocity head

z is the elevation head.

In the Bernoulli equation, hydraulic grade line (HGL) is the representation of the sum of static pressure and the elevation head, i.e., $\frac{P}{\rho g} + z$.

On the other hand, energy grade line (EGL) is the representation of the total head of the fluid, i.e., $\frac{P}{\rho g} + \frac{C^2}{2g} + z$.

The difference between the heights EGL and HGL, i.e., $\frac{C^2}{2g}$ represents the dynamic head.

Now, going back to Fig. 1.5, as the flow passes through several conduits, such as pipes and tunnels, there will be some head loss (h_L), which modifies the Bernoulli equation into the following form:

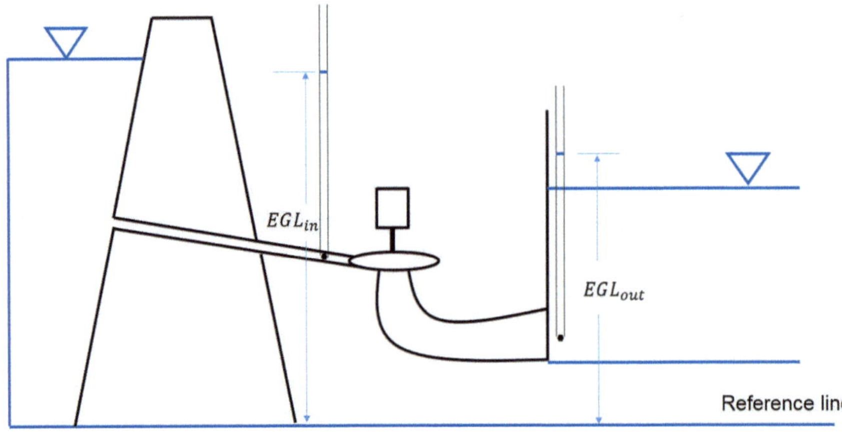

Fig. 1.5 Typical setup for a hydropower plant

$$\frac{P_1}{\rho g} + \frac{C_1^2}{2g} + z_1 = \frac{P_2}{\rho g} + \frac{C_2^2}{2g} + z_2 + h_L \tag{1.3}$$

where the subscripts 1 and 2 represent locations at inlet of the conduit and inlet of the turbine respectively.

Hence, the net head of a turbine is technically defined as the difference in the EGL upstream of the turbine and the EGL at the exit of the draft tube.

1.1.6 Efficiency of a Hydraulic Turbine

For calculation of the efficiency of a turbine, the ratio of the shaft power and the hydraulic power (power that is extracted by the turbine due to the water flowing) is taken into consideration.

$$\eta_{\text{turbine}} = \frac{\tau.\omega}{\rho.g.Q.H} \tag{1.4}$$

where

τ is the total torque produced by the runner

ρ is the density of the flow

Q is the designed discharge (volumetric flow rate in m^3/s) of the plant

H is the net head.

In the above equation, $\tau.\omega$ represents shaft power and $\rho.g.Q.H$ represents total power supplied to the turbine from water.

According to Euler's turbine equation developed by Leonhard Euler in the eighteenth century, output power from the runner can be written in the following form:

$$\dot{W} = \rho.Q.(U_1.C_{u1} - U_2.C_{u2}) \tag{1.5}$$

where the symbols in the term $(U_1.C_{u1} - U_2.C_{u2})$ represent different velocities and their components as described in the above section.

$\dot{W}$ is the power output from the runner. Hence, the efficiency related to the runner only is referred to as the hydraulic efficiency, η_{h}, and it is calculated as:

$$\eta_{\text{h}} = \frac{(U_1.C_{u1} - U_2.C_{u2})}{g.H} \tag{1.6}$$

1.1.7 Reduced Quantities

In the following chapters, velocity and flow parameters are converted into dimensionless forms in some cases, so that the parameters are independent of the size of the turbine.

For example:

$$\underline{C} = \frac{C}{\sqrt{2gH}} \tag{1.7}$$

The absolute velocity in the above equation is divided by the highest possible velocity which can be obtained if the net head is transferred into kinetic energy, i.e., $\sqrt{2gH}$.

It is to be noted that the parameters, ω and Q, can also be converted into reduced forms. However, the reduced values in these cases will not be dimensionless. The unit of $\underline{\omega}$ is $\left[\frac{1}{m}\right]$, and the unit of $\underline{Q}$ is $\left[m^2\right]$. Here, since the dimension is $\left[m^2\right]$, reduced flow can be considered as a measure of the cross section of the turbine outlet.

1.1.8 Francis Turbines

Turbines in hydropower plants can be classified depending upon their degree of reaction, which is the ratio of the pressure drop across the runner to the static pressure drop across the stage. In the case of impulse turbines, such as Pelton and Turgo, all the pressure head is converted into the dynamic head across the stationary components, whereas no pressure drop occurs inside the runner. In the case of reaction turbines, such as Francis and Kaplan, a proportion of the pressure drop occurs inside the runner, whereas rest occurs inside other stationary components. During the design of any power plant, the selection of the turbines is carried out depending upon the available head and discharge in the site. A typical turbine selection curve is shown in Fig. 1.6. Apart from the head and discharge, the speed of the turbine also plays a crucial role in determining its type. Hence, the selection of the most suitable turbine depends on a dimensionless number given by the following equation:

$$\Omega = \underline{\omega}.\sqrt{\underline{Q^*}} \tag{1.8}$$

In the above equation, Ω is called as speed number, which depends on the discharge, rotation speed, and head of the power plant. The quantities, $\underline{\omega}$ and $\underline{Q}^*$, with the under-bar are the reduced quantities and need to be divided by $\sqrt{2gH}$. As discussed in the previous chapter, since the unit of $\underline{Q}$ is $\left[m^2\right]$ and the unit of $\underline{\omega}$ is $\left[\frac{1}{m}\right]$, the above equation results in a dimensionless number. The asterisk sign after flow, i.e., $\underline{Q}^*$ is representing the flow in the designed condition. A rule of thumb is that power plants with speed number of less than 0.2 are suitable for Pelton turbines and

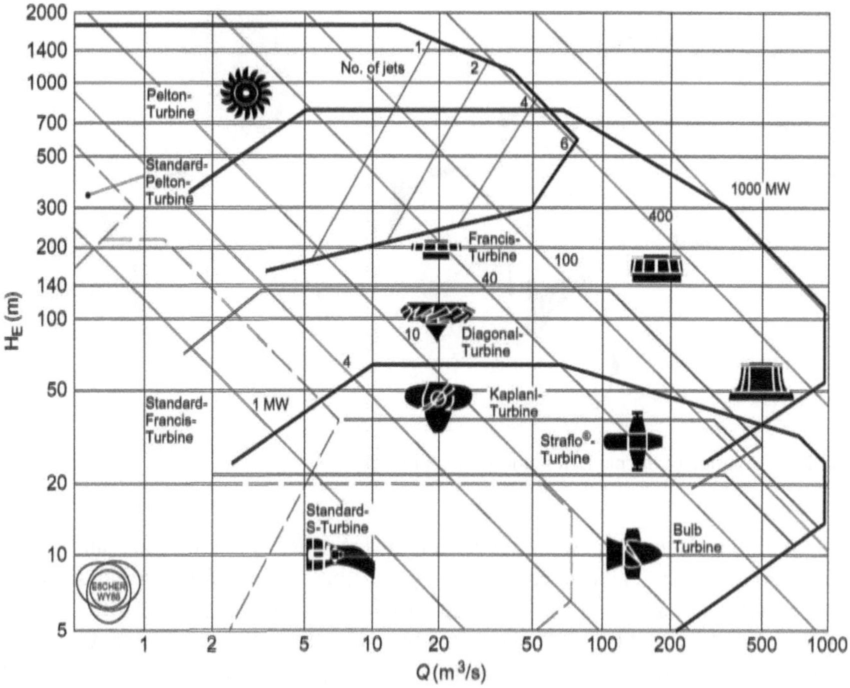

Fig. 1.6 Turbine selection curve (Retrieved from [2] with permission from Elsevier, Copyright © 2010 Elsevier)

those that are between 0.2 and 1.5 are suitable for Francis turbines. This suggests that Francis turbines have a very wide range of application, because of which the shape of the turbine varies, depending upon the speed number. The speed number close to 0.2 indicates high-head-type Francis turbines. They have runners that have small inlet height and long blades, with inlet diameter larger than the outlet diameter. As the speed number increases, the shape of the runner changes with high inlet height, short blades, and outlet diameter larger than the inlet diameter. An illustration of this transition is also shown in Fig. 1.6. This means that with higher value of the speed number, the shape of the runner changes from Francis type to Kaplan type.

The modern-day Francis turbines are designed and fabricated using the centuries of research works and experiences to reach the efficiency ranging more than 95%. It has been reported in one of the oldest literatures of mixed-flow turbines that the first Francis turbine was invented by James B. Francis together with Uriah A. Boyden in 1847, whose efficiency was measured to be 79.37% (computed efficiency was 79.31%). Due to the promising performance of this runner compared to previous turbines (or wheels), some rapid design optimizations took place for the next few decades to develop it as one of the most efficient energy conversion machines. The evolution of the Francis runner from 1847 is shown in Fig. 1.7. While initially the blades covered only a small portion of the overall area of the band/crown, the shape

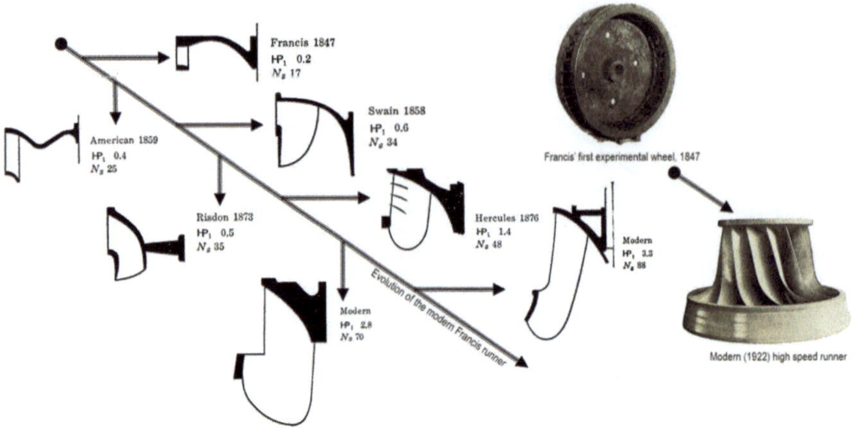

Fig. 1.7 Evolution of the modern Francis runner blades

was continuously revamped by several engineers, after series of trials and errors. After adopting the scientific method to design the turbines for specific sites, the Francis turbines started getting its modern runner appearance. By 1922, the Francis runner, also shown in Fig. 1.7, was already reported to give an efficiency of more than 90%, which was a remarkable achievement, typically for a low-head runner [3, 4]. In the current time, Francis turbine is the most used type of hydraulic turbines, harnessing more than half of the total energy generated from hydropower projects [5].

1.1.9 Components of Francis Turbines

A 3D model of a Francis turbine is shown in Fig. 1.8. The hydraulic design needs to be carried out for spiral casing, stay vanes, guide vanes, runner, draft tube, as well as labyrinth seals. The remaining part, such as shaft, bearing, head covers, and gate operating ring, are related to the mechanical design. In this book, we will be mostly focusing on the hydraulic part, apart from some cases, such as the thickness of the runner, where basic mechanical design principles are incorporated.

The components used in Francis turbines along with their role are briefly described below:

Spiral casing

The main role of spiral casing is to transfer water from upstream into the runner through stay vanes and guide vanes. The decreasing cross section maintains constant flow velocity at the inlet of the stay vanes, and water is distributed uniformly.

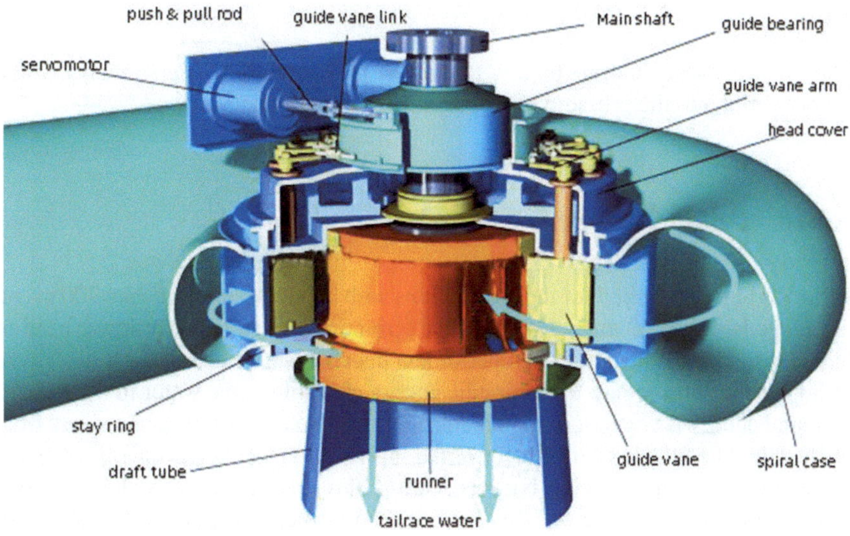

Fig. 1.8 Model of Francis turbine [6]

Stay Vanes

The stay vanes add structural support to the spiral casing and prevent it from tearing apart. These vanes are shaped to have a minimum influence on the flow. However, there is a small increment in the tangential component of absolute velocity, C_u as the design follows a free vortex theory.

Guide vanes

The main purpose of guide vanes is to regulate the flow into the turbine. Guide vanes are designed so that they are able to rotate, depending upon the operating condition. The guide vane opening angle corresponding to the designed discharge is called as best efficiency point (BEP), because the turbines are designed to achieve the highest efficiency at this operating condition.

Runner

It is considered as the heart of the turbine, where the hydraulic energy from water is converted into useful forms. The runner is connected to the shaft, and hydraulically, it is the only component that is rotating. Hence, the concept of relative velocity is valid only inside the runner.

Labyrinth seals

Labyrinths help to minimize the leakage loss between the turbine runner and the cover, as well as balance the axial forces in the turbine. It consists of a static seal connected to the covers and a rotating part connected to the runner.

Draft tube

The main purpose of draft tube is to convert the remaining kinetic energy at the runner outlet into the pressure energy at the draft tube outlet.

1.1.10 Pelton Turbines

Pelton turbines are suited for the applications with high head of more than 700 m and discharge less than 60 m^3/s. However, the turbine is also suitable for lower head or higher discharge, if the speed number of less than 0.2 can be maintained. It is a type of impulse turbine, where the hydraulic energy available in the water in the form of potential energy is converted into kinetic energy in the form of high-speed jets from the nozzle. The kinetic energy is converted further into mechanical energy when the jet interacts with the rotating buckets. During this interaction, only the impulsive force of the flow is effective. This force on the bucket (F_{bucket}) can be calculated based on the following equation:

$$F_{\text{bucket}} = \dot{m}_c \times \frac{w^2}{C_0}(1 - \cos \beta_2) \quad (1.9)$$

The symbol C_0 represents the absolute velocity of the flow at the inlet of the bucket, and β_2 represents the outlet angle, as shown in Fig. 1.9a. $\vec{W} = \vec{C} - \vec{U}$ is the relative velocity of the jet with respect to the bucket, which is assumed to be constant at the inlet and the outlet of the stream. The power received by the bucket can then be calculated by multiplying Eq. 1.1 with the rotational speed of the turbine.

In Eq. 1.1, the total mass flow rate of the jet is defined as,

$$\dot{m}_c = \rho \times C_0 \times A_{jet} \quad (1.10)$$

where ρ is the density of the fluid and A_{jet} is the cross-sectional area of the jet.

The history of turbine's development starting from water wheels of early nineteenth century was reviewed by Viollet [7], which showed a progressive development in the turbine's unit power, range of operations, and efficiency along the years. Although the concept of water wheel was developed in early 1800s, the concept of the flow into two halves of the bucket was patented by Lester Pelton in 1880s [8]. This design gave the possibility of flow escaping smoothly, utilizing the reactionary force of the escaping stream in addition to the direct force of the impinging jet. The concept of Pelton turbine was modified for higher specific speed by Eric Crewdson and developed the first Turgo impulse turbine in 1919 by placing the angle of impact of the jet and the wheel plane at 20° and the inlet edge peripheral velocity 0.45–0.47 of the ideal jet velocity [9]. The evolution of turbines based on the unit power is shown in Fig. 1.9b [7]. Before the development of Francis and Pelton turbines, Fourneyron and Girard turbines were popular, mostly in the milling industries. However,

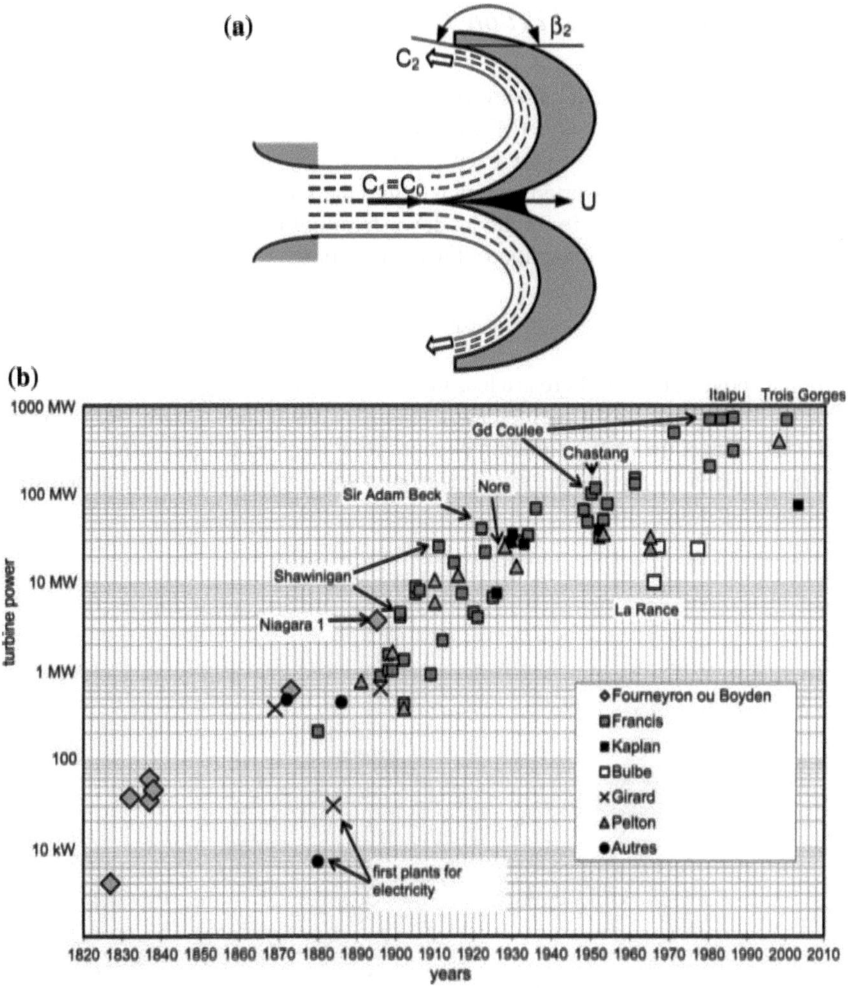

Fig. 1.9 **a** Flow behavior within the nozzle and the bucket and **b** turbines' unit power from 1825 (Retrieved from [7] with permission from Elsevier Masson, Copyright © 2017 Elsevier Masson SAS)

during the twentieth century, both Francis and Pelton turbines flourished the market gradually from few megawatts to several hundred megawatts per unit. At present, the largest unit of Francis turbine that is installed has the capacity of 700 MW at both Three Gorges and Itaipu hydropower plants. Similarly, the largest unit of Pelton turbine is installed at Bieudron hydropower plant with the rated capacity of 423 MW and with a head of 1869 m. Viollet [7] also reported that with the development of wind and solar energies in recent years, the hydropower plants are facing the challenges of flexibility due to operations in unsteady regimes.

1.1.11 Components of Pelton Turbines

Figure 1.10 shows the sectional view of a multi-jet Pelton turbine. The main components after the main inlet valve (MIV) of such turbines consist of manifold, nozzle, spear valves, and buckets, apart from other auxiliary components. The potential energy from water is converted completely into the kinetic energy in the nozzle, as a result of which the velocity of the jet downstream of the nozzle can be calculated for an ideal condition by using the following equation:

$$v_{jet(ideal)} = \sqrt{2.g.H} \tag{1.11}$$

In the above equation, there is a loss in the velocity due to the shape of the nozzle, which is incorporated by a coefficient of velocity, Cv which is usually between 0.95 and 0.99. Hence, the above equation changes to: $v_{jet(real)} = C_v.\sqrt{2.g.H}$.

The diameter of the jet that emerges from a spear valve can be calculated from the following equation:

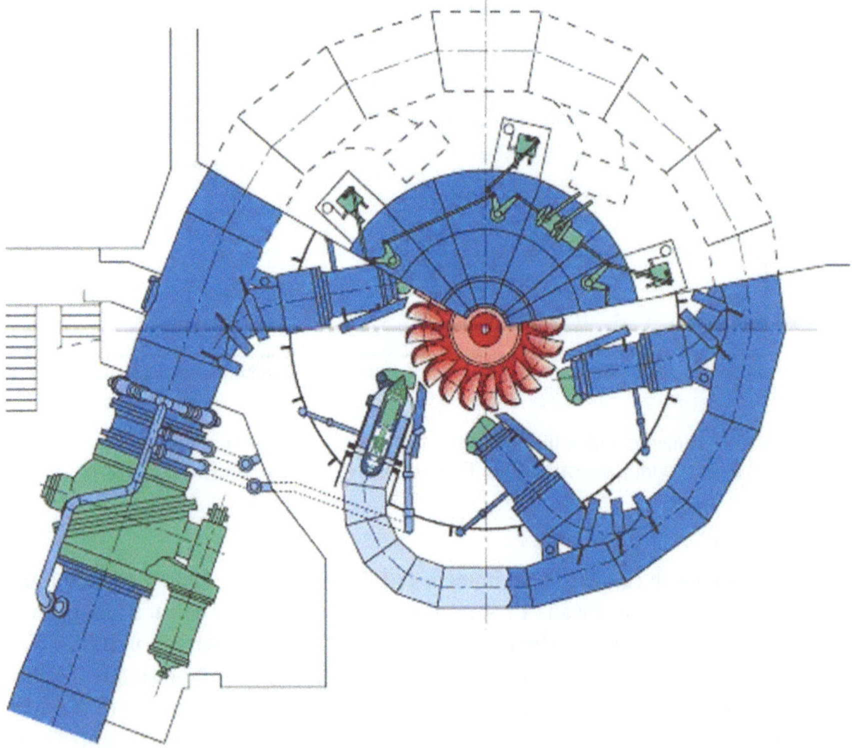

Fig. 1.10 Sectional view of Pelton wheel [10]

$$d_{\text{jet}} = \sqrt{2.\sin\frac{\alpha}{2}(2.s.d_{\text{noz}} - s^2.\sin\alpha)} \tag{1.12}$$

where

α = spear angle, for example 55°

s = spear travel (same unit as d_{jet}).

The above equation can be used if the main dimensions of the nozzle and spear are known. Alternatively, the diameter of the jet during an initial design phase can be calculated based on the equation of continuity, i.e.,

$$Q = z_n \cdot \frac{\pi \cdot d_{jet}^2}{4} \cdot v_{jet} \tag{1.13}$$

where

z_n = number of nozzles.

1.2 Sediment Erosion

Hydropower plants in the regions of Himalayas and Andes are prone to excessive sedimentation issues, which may cause damage of the turbine components. Although many types of wear exist in hydraulic machineries, erosion due to sediments and cavitation are the types of wear which are relevant to the context of sediment laden hydropower turbines. Quantifying the erosion in turbines has been one of the major research challenges because erosion is affected by several factors, such as flow behavior, properties of the sediments, as well as properties of the eroded materials. Erosion models found in literature suggest that the most general expression of erosion is Erosion $\propto$ (Velocity)^n, where the value of n depends on material and operating conditions. In the case of hydraulic turbines, the flow is highly turbulent, which accelerates the erosion and causes loss in efficiency and decrease in the turbine's lifespan.

In recent times, the importance of using computational fluid dynamics for predicting erosion in hydraulic turbines has been discussed in literatures. Nevertheless, the use of OpenFOAM for such analyses is a relatively new topic. In Chaps. 5–7, methods of estimating the erosion in Francis runner and Pelton nozzle have been discussed using OpenFOAM.

References

1. Y.A. Cengel, J.M. Cimbala, *Fluid Mechanics, Fundamentals and Applications* (Mc Graw Hill, New York, 2006)

2. S.L. Dixon, C.A. Hall, *Chapter 9—Hydraulic Turbines, Fluid Mechanics and Thermodynamics of Turbomachinery* (Butterworth-Heinemann, 2014)
3. B.J. Lewis, J.M. Cimbala, A.M. Wouden, Major historical developments in the design of water wheels and Francis hydroturbines. IOP Conf. Ser. Earth Environ. Sci. **22**, 012020 (2014)
4. A.T. Safford, E.P. Hamilton, The American mixed-flow turbine and its setting. Trans. Am. Soc. Civil Eng. **85** (1922)
5. C. Trivedi, I. Iliev, O.G. Dahlhaug, Numerical study of a Francis turbine over wide operating range: some practical aspects of verification. Sustain **12**(10), 4301 (2020)
6. Huahydro. https://www.huahydro.com/product/francis-turbine-runner/. Retrieved on August, 2023, CC BY-SA 3.0
7. P. Viollet, From the water wheel to turbines and hydroelectricity. Technological evolution and revolutions. Comptes Rendus Mecanique **345**, 570–580 (2017)
8. L.A. Pelton, Water wheel. US Patent, No. 409,865 (1889)
9. D.S. Benzon, The Turgo impulse turbine; a CFD based approach to the design improvement with experimental validation. PhD thesis. Lancaster University (2016)
10. Voith Siemens Hydro Power Generation. http://www.voithsiemens.com. CC BY-SA 3.0

Chapter 2
Calculation of the Main Dimensions of Francis Runner

The design process of Francis turbines starts with calculating the main dimensions of the runner, i.e., outlet diameter, D_2, inlet diameter, D_1, inlet height, B_1, and components of the velocity triangles at inlet and outlet. These dimensions are shown in Fig. 2.1. The required inputs from the power plant are the design head, H, and the flow rate, Q.

Before moving ahead with the design, a small example containing the revision of Chap. 1 is done, to understand the velocity triangles, specifically for Francis turbines. A schematic of guide vanes and runner is shown in Fig. 2.2, together with the velocity triangles at inlet and outlet. As it is understood now that the relative velocity components are added once the flow reaches the runner inlet. The velocity of the runner, i.e., the tangential velocity at the inlet, U_1, is 61.9 m/s (represented by the blue line), whereas at the outlet, U_2, is 40 m/s. Therefore, the runner in this case is representing a high-head type, with $D_1 > D_2$. The absolute velocities are represented by red solid lines, and their components in "u" and "m" directions are represented by the dash red lines. The relative velocities are represented by blue lines. The other components that are not shown in the figure can be calculated using Pythagoras. The summary of the parameters in Fig. 2.2 is listed in Table 2.1. Note that the subscripts 1 and 2 are representing the inlet and the outlet of the runner blade respectively.

2.1 Dimensions at the Runner Outlet

As it could be noticed from Fig. 2.3, the tangential component of absolute velocity at the outlet is 0. This is the assumption made to avoid any rotational motion of the flow at the outlet of the runner. However, this rotational motion cannot be avoided in the part load and full load operation. Figure 2.3 shows different possibilities of the velocity triangle at the outlet of the runner. In both part load and full load conditions, the direction of absolute velocity is not perpendicular to the tangential velocity. In

Z. Guo et al., *Design of Francis Turbine and OpenFOAM Methods*,
https://doi.org/10.1007/978-981-99-8381-0_2

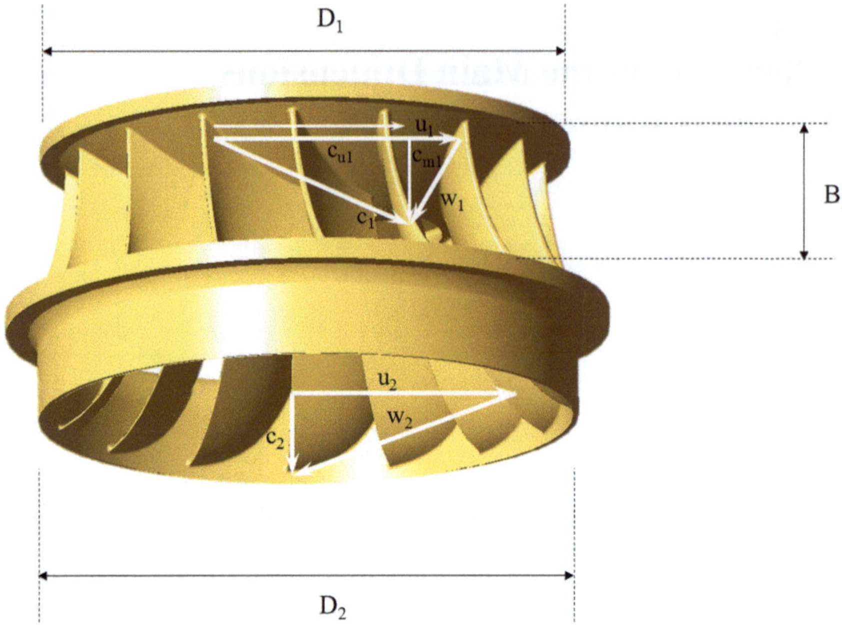

Fig. 2.1 Main dimensions of Francis runner

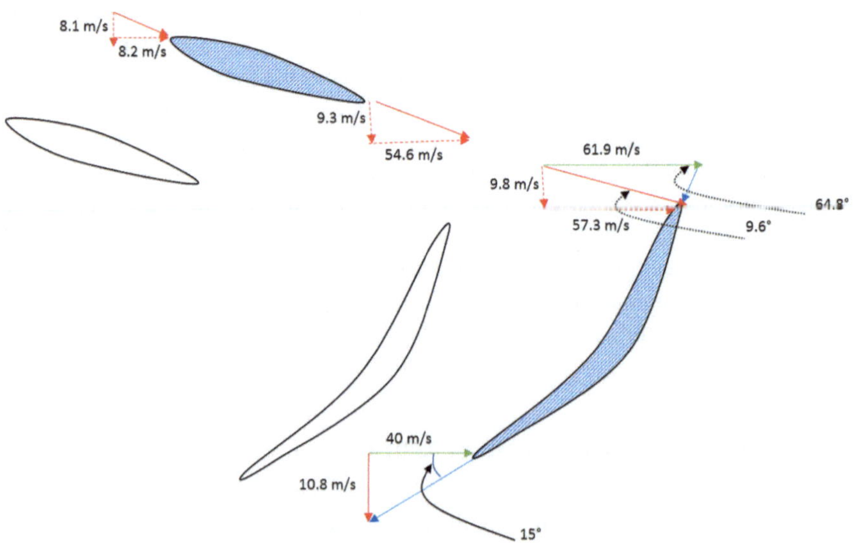

Fig. 2.2 Example of velocity triangles in Francis turbines (guide vanes and runner)

Table 2.1 Magnitude of the parameters in velocity triangles shown in Fig. 2.2

Parameter	Value	Parameter	Value
C_{u1}	57.3 m/s	W_1	10.82 m/s
C_{m1}	9.8 m/s	W_2	41.43 m/s
C_1	58.13 m/s	β_2	15°
U_1	61.9 m/s	α_2	90°
β_1	64.8°	C_{u2}	0 m/s
α_1	9.6°	C_{m2}	10.82 m/s

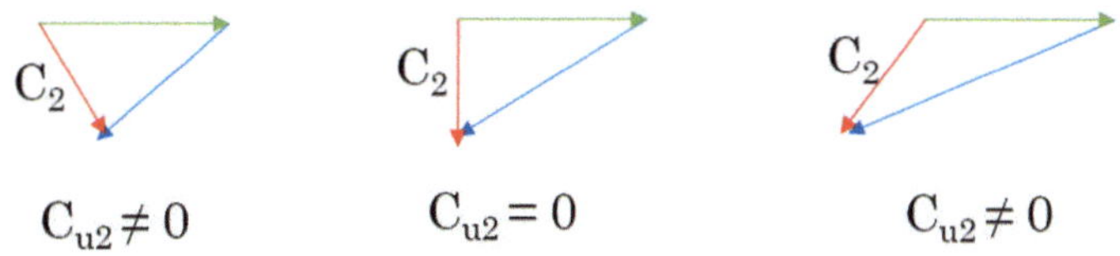

Fig. 2.3 Possible velocity triangles at the outlet

the case of part load operation, the direction of C_{u2} is same as the direction of U_2, whereas it is opposite for the case of full load operation. As a result, a swirl motion of the flow can be seen in the draft tube having various shapes depending upon the operating conditions. However, for the design of the runner, the BEP condition is taken without any swirl component.

$$C_{u2} = 0 \tag{2.1}$$

For the selection of β_2 and U_2, empirical data are referred depending upon the past experiences.

$13° < \beta_2 < 22°$ Lowest value for highest head.
$35\,\text{m/s} < U_2 < 42\,\text{m/s}$ Highest value for highest head.

Higher value of U_2 is found to have increased noise and vibrations in the turbine. When these values are selected, rest of the parameters at the outlet can be calculated. From the right-angled triangle at the outlet:

$$C_{m2} = U_2.\tan\beta_2 \tag{2.2}$$

From the definition of flow rate, the outlet diameter of the runner can be calculated. In this case, the velocity responsible for the flow is C_{m2}, since the tangential component determines the power produced from the turbine.

$$D_2 = \sqrt{\frac{4Q}{\pi C_{m2}}} \tag{2.3}$$

Similarly, from the definition of the rotation speed in rpm:

$$n = \frac{U_2.60}{\pi D_2} \tag{2.4}$$

It is important to calculate the number of the poles in the generator at this stage to know the closest synchronous speed to match the net frequency of a particular country. The number of poles is obtained by using the relation:

$$Z_{\text{poles}} = 60.\frac{f}{n} \tag{2.5}$$

For example, if the net frequency of the country is 50 Hz, f in the above equation is replaced with 50 Hz. Note that the above equation is dimensionless, since the minute is already converted into seconds by multiplying with 60. In the case of the generator with 3000 rpm, above equation yields, $Z_{\text{poles}} = 1$.

As the number of poles must be an integer, the value obtained from the above equation is rounded up. As a result, the new values of D_2 and n need to be calculated. From Eq. 2.2, keeping the angle, β_2, as constant,

$$\tan\beta_2 = \frac{C_{m2}}{U_2} = \frac{C_{m2\text{corrected}}}{U_{2\text{corrected}}} = \frac{\frac{4Q}{\pi D_2^2}}{\frac{\pi n D_2}{60}} = \frac{\frac{4Q}{\pi D_{2\text{corrected}}^2}}{\frac{\pi n_{\text{corrected}} D_{2\text{corrected}}}{60}}$$

$$D_{2\text{corrected}} = \sqrt[3]{\frac{n D_2^3}{n_{\text{corrected}}}} \tag{2.6}$$

2.2 Submergence

The level of the turbine should be sufficiently lower than the tail race level to avoid cavitation at the outlet of the runner. A schematic of a turbine is shown in Fig. 2.4. Two points, one at the outlet of the runner and another at the outlet of the draft tube, are taken into consideration. Taking the static, dynamic, and elevation head at these two locations and applying the Bernoulli's equation:

$$h_2 + \frac{c_2^2}{2g} + Z_2 = h_3 + \frac{c_3^2}{2g} + Z_3 + J_3 \tag{2.7}$$

In Eq. 2.7, $J_3 = \frac{\zeta_3 c_3^2}{2g}$ is representing the friction in the draft tube and bend loss. Also, h_2 is the absolute pressure, while h_2 includes the barometric pressure h_b. This gives,

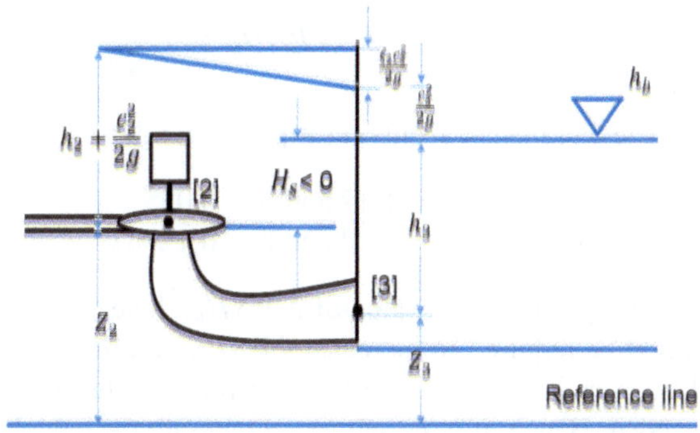

Fig. 2.4 Layout of a Francis turbine indicating different heads

$$h_3 = Z_2 - Z_3 - H_s + h_b \tag{2.8}$$

In Eq. 2.8, H_s represents the suction head, and it has a negative value if point 2 is below the water level outside the draft tube.

From Eqs. 2.7 and 2.8,

$$h_2 = -H_s + h_b + \frac{c_3^2}{2g} + J_3 - \frac{c_2^2}{2g} \tag{2.9}$$

It can be inferred from Eq. 2.9 that higher value of c_2 gives lower value of h_2. If the value of h_2 is far below the atmospheric pressure, the cavitation might occur, which means that the boiling point of water will be achieved at the normal room temperature. In order to avoid cavitation, the suction head, H_s, can be increased, i.e., the turbine can be further placed lower than the tail race level. If Eq. 2.10 is satisfied, the cavitation in turbines can be avoided:

$$h_2 = -H_s + h_b - \left(\frac{c_2^2}{2g} - \frac{c_3^2}{2g} - J_3\right) > h_{\text{vapor}} \tag{2.10}$$

where h_{vapor} is the vapor pressure of water. Rearranging Eq. 2.10, $-H_s + h_b - h_{\text{vapor}}$ represents the head that is available in the power plant, also called as net positive suction head available, $NPSH_A$. Hence, the requirement mentioned in Eq. 2.10 can be rewritten as:

$$-H_s + h_b - h_{\text{vapor}} = NPSH_A > \frac{c_2^2}{2g} - \frac{c_3^2}{2g} - J_3 \tag{2.11}$$

The terms in the right-hand side of Eq. 2.11 represents the NPSH that is required from the turbine, also called as net position suction head required, $NPSH_R$, i.e., $\text{NPSH}_R = \frac{c_2^2}{2g} - \frac{c_3^2}{2g} - J_3$. From the perspective of the design of Francis turbines, some empirical formulas exist that can be used to calculate the required suction head. One of them is provided in Eq. 2.12.

$$NPSH_R = a.\frac{C_{m2}^2}{2g} + b.\frac{U_2^2}{2g} \tag{2.12}$$

The parameters a and b in Eq. 2.12 are dependent on the speed number:

$\Omega < 0.55$ gives a = 1.12 and b = 0.055.
$\Omega > 0.55$ gives a = 1.12 and b = $0.1 \times \Omega$.

From Eq. 2.12, if the value of $NPSH_R$ comes out to be too large, high value of H_s might be required to increase $NPSH_A$ and minimize the possibility of cavitation. However, it might not be feasible for the site to provide the suction head that is needed. Hence, it is necessary to reiterate the calculation of the dimensions at the outlet using a lower value of U_2.

2.3 Dimensions at the Runner Inlet

Dimensioning the inlet starts with the Euler equation, which was discussed in Chap. 1.

$$\eta_h = \frac{(U_1.C_{u1} - U_2.C_{u2})}{g.H} \tag{2.13}$$

Since $C_{u2} = 0$ and introducing the reduced terms in Eq. 2.13:

$$\eta_h = 2.\underline{U_1}.\underline{C_{u1}} \tag{2.14}$$

The speed number of the turbine was defined in Sect. 1.5. By the definition of $\underline{\omega}$ and $\underline{Q}^*$, the speed number can be rewritten as:

$$\Omega = \frac{U_1}{D_1} \cdot \frac{1}{\sqrt{2gH}} \cdot \sqrt{\pi D_2^2 \cdot \frac{C_2}{\sqrt{2gH}}} = \frac{D_2}{D_1} \cdot \underline{U_1} \cdot \sqrt{\pi \cdot \frac{C_2}{\sqrt{2gH}}} \tag{2.15}$$

It can be seen from Eq. 2.15 that with the increasing speed number of the turbine, the peripheral velocity U_1 should also increase, which consequently decreases the values of β_1 and β_2 by keeping C_2 constant. This also eventually decreases the value of U_2, as illustrated in Fig. 2.5.

It can also be seen from Eq. 2.15 that with the increasing speed number, the ratio $\frac{D_2}{D_1}$ must also increase. This means that the diameter D_2 keeps increasing for turbines with high speed number, and the runner takes the shape as shown in Fig. 2.5. As the value of Ω is also dependent on C_2, there cavitation effect could be more predominant in such high speed number turbines.

The reduced peripheral velocity at the inlet can be chosen based on experience. Its value ranges from 0.7 to 1 depending upon the speed number of 0.2 to 1. For high-head Francis turbines, this is typically ranging from 0.71 to 0.73. By assuming

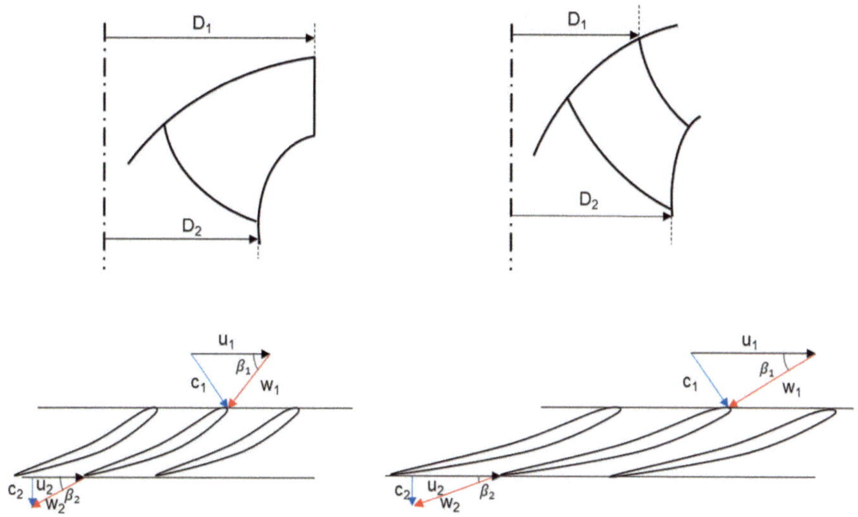

Fig. 2.5 Scenarios for high and low speed number turbines

the efficiency, η_h as 0.96, the value of $\underline{C_{u1}}$ can be calculated from Eq. 2.13. The reduced values can be converted into the actual value by multiplying with $\sqrt{2gH}$.

$$U_1 = \underline{U_1}.\sqrt{2gH} \tag{2.16}$$

The inlet diameter can be calculated using:

$$D_1 = \frac{U_1.60}{\pi n} \tag{2.17}$$

Similarly, the inlet blade angle can be calculated from the velocity triangle at the inlet:

$$\tan \beta_1 = \frac{C_{m1}}{U_1 - C_{u1}} \tag{2.18}$$

In Eq. 2.17, the value of C_{m1} is calculated based on the desired acceleration of the flow from inlet to outlet, so that the backflow in the runner can be prevented. For example, if the desired acceleration is 10%, the following condition can be used:

$$C_{m2} = 1.1 \times C_{m1} \tag{2.19}$$

Using the continuity equation, the inlet height of the runner can be calculated:

$$A_1.C_{m1} = A_2.C_{m2} \tag{2.20}$$

It is to be noted from Eq. 2.19 that the area A_1 represents the total flow passage at the inlet, which is equivalent to the curved surface area of a cylinder with height B_1 and diameter D_1. On the other hand, the area A_2 represents the total flow passage at the outlet, which is equivalent to the area of a circle with diameter D_2. Hence, $A_1 = \pi . D_1 . B_1$ and $A_2 = \pi . \frac{D_2^2}{4}$. Putting these values in Eq. 2.19 and using the relation 2.18:

$$B_1 = \frac{1.1 D_2^2}{4 D_1} \tag{2.21}$$

Chapter 3
Shape of the Runner Blades

This chapter focuses on providing a hydraulic shape of the runner blades. The methods used in this chapter are typically suitable for high head type of Francis turbines, where $D1 > D2$. At first, the axial view of the blade is constructed, followed by the radial view. Step-wise process of designing the runner blades is provided in the following sections.

3.1 Axial View

The first streamline, either hub or shroud, is considered to be following a circular or elliptical profile. The general equation of an ellipse with a and b as major and minor axes can be represented by Eq. 3.1.

$$\frac{x^2}{a^2} + \frac{y^2}{b^2} = 1 \tag{3.1}$$

If the ellipse is divided in four quadrants, with center at the origin, either first or second quadrant is suitable for developing the axial view. In this case, as shown in Fig. 3.1, the x co-ordinate belongs to the second quadrant, i.e., from 0 to -1. In this case, the major axis, i.e., a is 1 and the minor axis, i.e., b is 6. These values can be re-iterated if the change in the design is preferred. In the following example, the first curve is representing the shroud, however, it is also possible to start with the hub.

For practicality, the curve shown in Fig. 3.1 needs to be translated in the correct position. Here, since the axis of rotation needs to pass from the origin, the curve was translated as shown in Fig. 3.2. Moreover, the shroud should end at $D2/2$, hence, additional points after the radius of the outlet are trimmed off. If any programming tool is being used for developing the axial view, it is also practical to redistribute the final section into a fixed desirable number of points through interpolation, so that it

Z. Guo et al., *Design of Francis Turbine and OpenFOAM Methods*,
https://doi.org/10.1007/978-981-99-8381-0_3

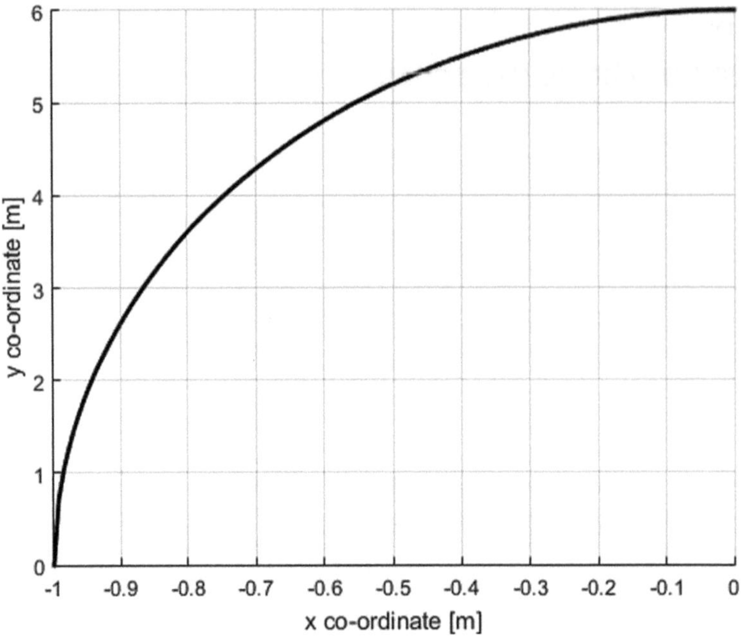

Fig. 3.1 First streamline of shroud made from the equation of an ellipse

will be easy to handle further steps. For example, 100 points for each curve provide a smooth line.

If "*i*" is representing number of points for each curve, "*j*" represents the number of streamlines. An equal flow for each streamline is assumed, so that the starting point of each streamline can be obtained using the known values, $B1$ and Q.

Similarly, as shown in Eq. 18, the acceleration inside the runner can be assumed as 10%. By using Eq. 3.2, the end points can also be obtained.

$$A_1 = 1.1 \times A_2 \tag{3.2}$$

When the start and the end points are known, rest of the points in each streamline is constructed. Geometrically, it is done by drawing a line between the points $(i + 1, j)$, (i, j), and $(i - 1, j)$. The new point, $(i, j + 1)$ is placed on an axis orthogonal to this line. Figure 3.3 shows the process of calculating the location of the new points in the neighboring streamlines.

From Figs. 3.2 and 3.3, following equations can be used to define all the streamlines. It is to be noted that a co-ordinate system $R - Z$ is used, which is representing a meridional plane.

$$\alpha_{i,1} = \tan^{-1}\left(\frac{Z_{i-1,1} - Z_{i+1,1}}{R_{i-1,1} - R_{i+1,1}}\right) \tag{3.3}$$

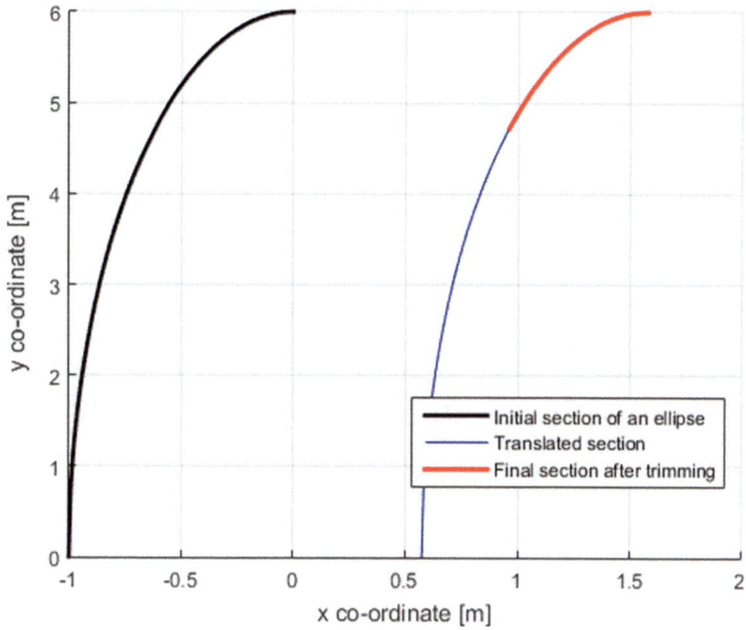

Fig. 3.2 Streamline of shroud after translation and trimming

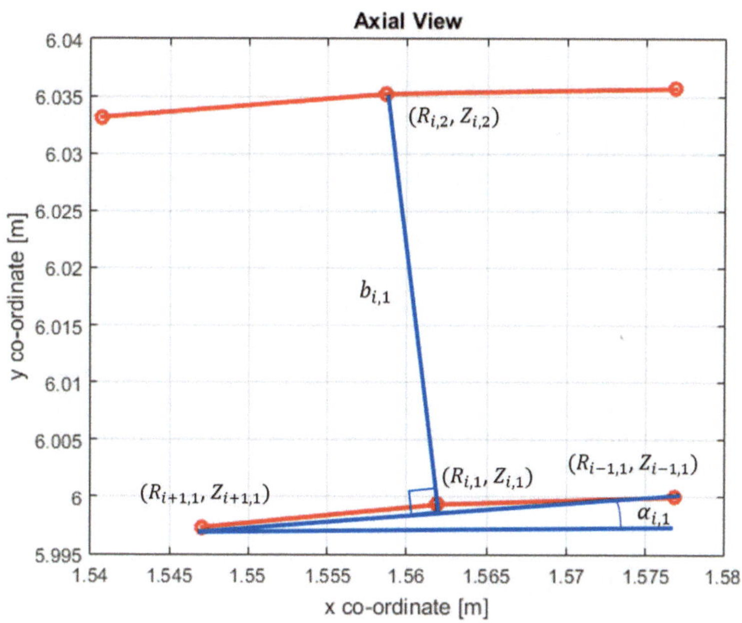

Fig. 3.3 Illustration of the process to find new points in adjacent streamlines

$$r_{l,1} = \frac{R_{l,1} + R_{l,2}}{2} \tag{3.4}$$

In Eq. 3.4, r is representing the distance of the mid-point of the height b from the center. The height is given by the relation:

$$b_{l,1} = \frac{R_{l,2} - R_{l,1}}{\sin \alpha_{l,1}} \tag{3.5}$$

The area of the curved surface area of a cylinder at the particular radius from Eqs. 3.4 and 3.5 is given by:

$$A_{l,1} = 2\pi \left(\frac{R_{l,1} + R_{l,2}}{2} \right) \left(\frac{R_{l,2} - R_{l,1}}{\sin \alpha_{l,1}} \right) = \frac{\pi}{\sin \alpha_{l,1}} \left(R_{l,2}^2 - R_{l,1}^2 \right) \tag{3.6}$$

Rearranging Eq. 3.6:

$$R_{l,2} = \sqrt{R_{l,1}^2 - \frac{A_{l,1} \sin \alpha_{l,1}}{\pi}} \tag{3.7}$$

Similarly, from Fig. 3.3:

$$Z_{l,2} = Z_{l,1} - b_{l,1} \cos \sin \alpha_{l,1} \tag{3.8}$$

Equations 3.7 and 3.8 provide the point in the neighboring streamline. In Eq. 3.7, $A_{l,1}$ can be calculated from the same method as shown in Eq. 3.2, and the distribution of the area from inlet to outlet can be considered as linear.

The same method can be followed for other points of the streamline or the adjacent streamlines. Figure 3.4 shows the axial view that will be obtained if the abovementioned equations are used in a programming tool. The figure shows that the curves around the hub are stretched, and they are nearly touching the center. As the turbine shaft will be connected at this location, and in order to avoid long blades, the end points around the hub region are trimmed off. For example, in this case, a sequence of 3*number of streamlines is used as the number of points from the trailing edge to be removed, so that 6 points are removed from the second streamline, 9 points from the second streamline, and so on. The final axial view obtained from this additional adjustment is shown in Fig. 3.5.

3.2 Energy Distribution

Before moving toward the radial view, it is important to choose the β distribution inside the runner, which not only defines the shape of the blade, but is also related to the energy distribution through the runner cascade. The energy distribution, from

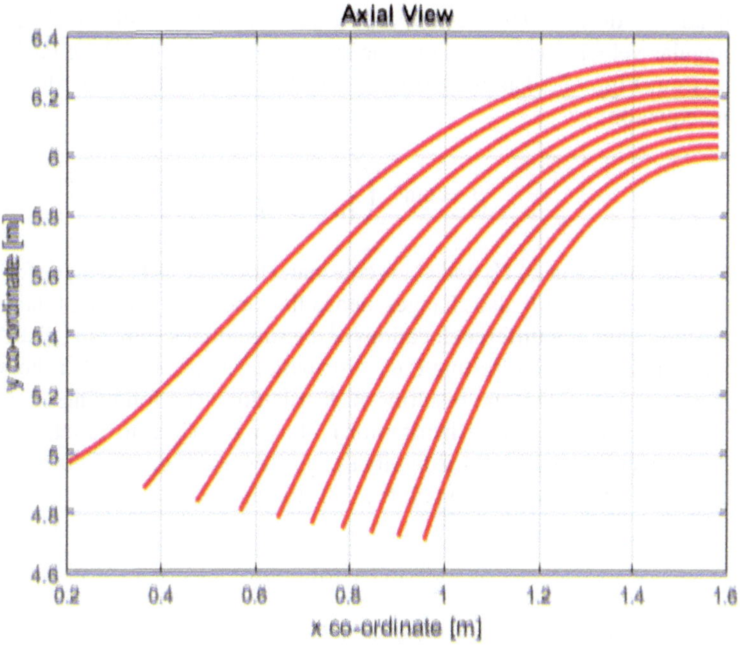

Fig. 3.4 Axial view obtained from the set of Eqs. 3.3–3.8

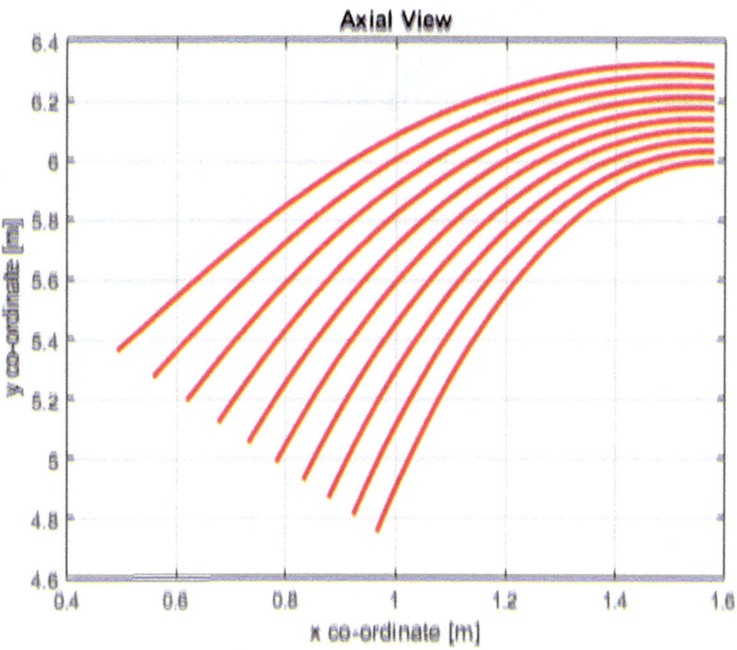

Fig. 3.5 Final axial view obtained after cutting the end points

Euler equation, is described by the product, UC_u. The peripheral velocity, U, depends on the angular velocity and the radius, hence, it can be calculated for each of the locations from Fig. 3.5. β is related to U and C_u at every locations through the following equation:

$$\beta = \tan^{-1}\left(\frac{C_m}{U - C_u}\right) \tag{3.9}$$

At the outlet, the value of C_u and C_m is constant. However, as U changes for all the streamlines, the values of β also changes, depending upon the location of the trailing edge from the center. For different locations of the streamlines, either the distribution of β can be chosen and the resulting UC_u distribution can be obtained or the reverse can be performed to obtain the resulting β distribution. The following example is shown for a case where a linear β distribution is chosen from inlet to outlet, illustrated in Fig. 3.6, and the resulting energy distribution is shown in Fig. 3.7. At this moment, the value of β is known for all the points in all streamlines.

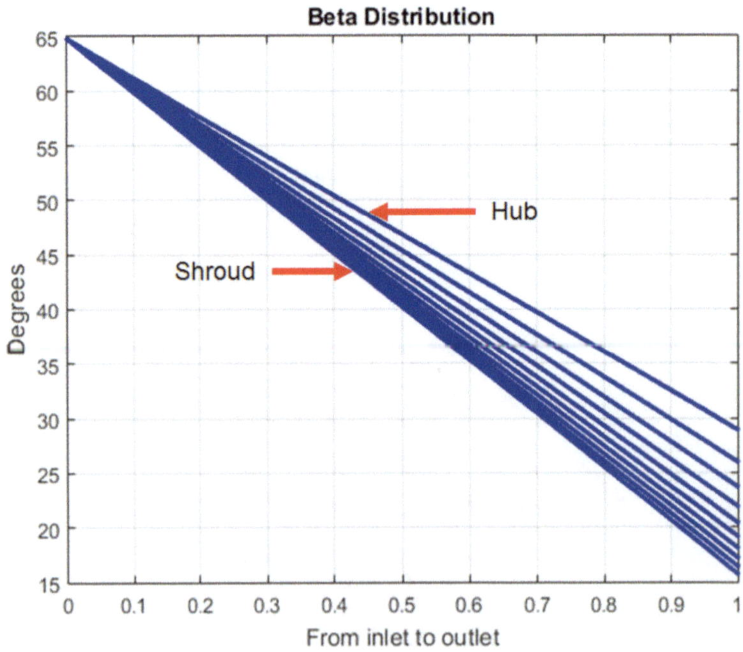

Fig. 3.6 Linear β distribution chosen from inlet to outlet for all the streamlines

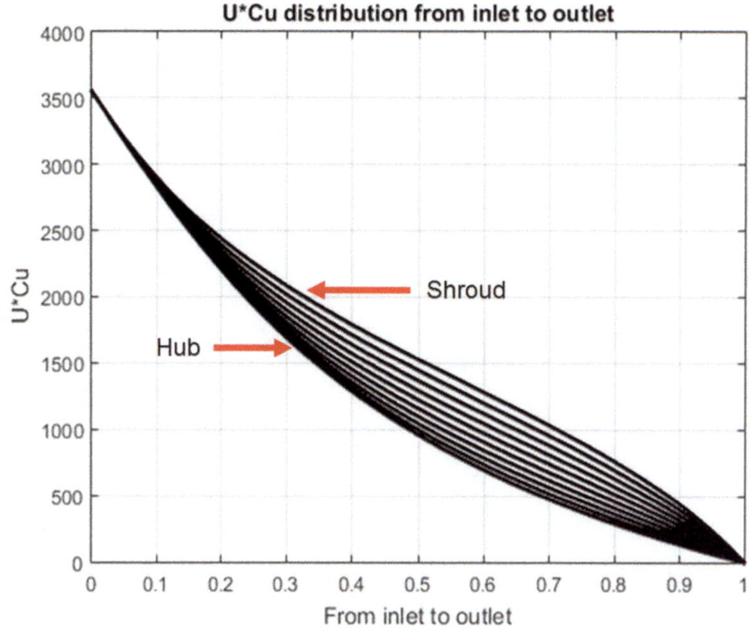

Fig. 3.7 Resulting UC_u obtained from inlet to outlet by assuming linear β distribution

3.3 Radial View

In order to construct the radial view of the runner, a new plane $G - H$ is defined. G is the length of a streamline in the axial plane, whereas H is the length of a streamline in the radial plane. By obtaining a relation between G and H, it will be possible to convert the axial view into the radial view.

From Fig. 3.3, for two adjacent points in a streamline, G can be calculated using Eq. 3.10.

$$G_{i,1} = G_{i-1,1} + \sqrt{\left(R_{i-1,1} - R_{i,1}\right)^2 + \left(Z_{i-1,1} - Z_{i,1}\right)^2} \tag{3.10}$$

The length of the streamline, G, can be divided into equal parts, ΔG. Since the β distribution is known from Sect. 3.2, the small length ΔH can be calculated using 3.11.

$$\Delta H = \frac{\Delta G}{\tan \beta} \tag{3.11}$$

After ΔG and ΔH are known, the radial view is constructed in polar co-ordinates, using the notation:

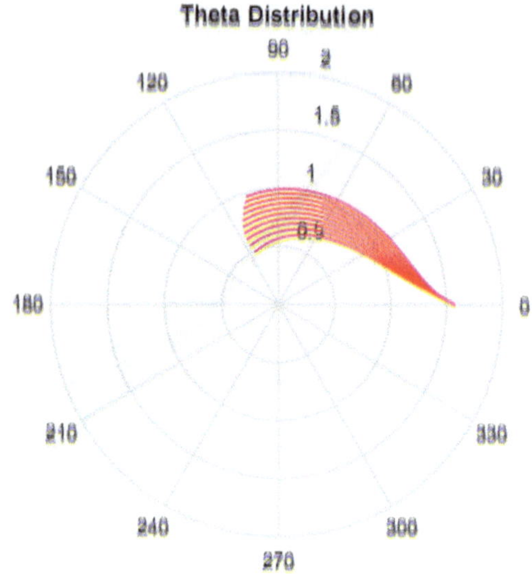

Fig. 3.8 Radial view of the runner in polar co-ordinates

$$d\theta = \frac{\Delta H}{R} \tag{3.12}$$

The resulting radial view obtained after obtaining the co-ordinates for θ and R is shown in Fig. 3.8.

Finally, the 3D shape of the blades can be obtained by using the axial and radial view of the runner. In Fig. 3.9, the 3D surface of a runner blade is shown, which is patterned around the axis, to obtain the full view of the runner blades, as shown in Fig. 3.10. It is to be noted that the blades are only a surface, in which a suitable thickness needs to be added to obtain the full 3D model.

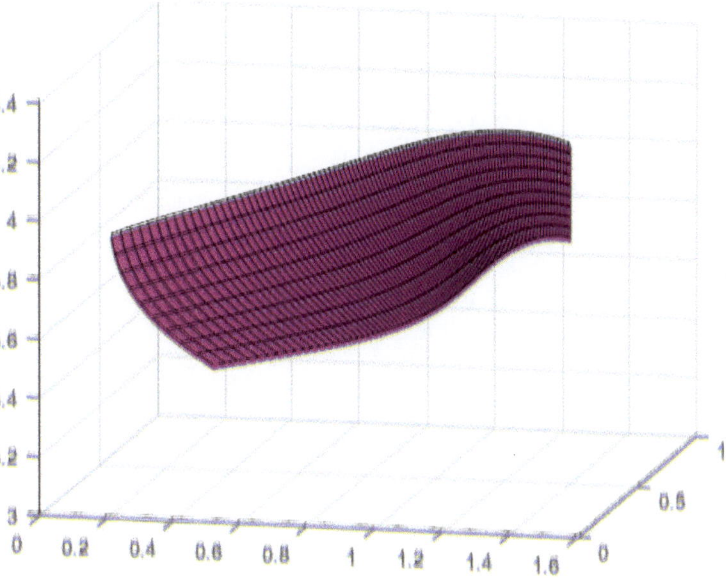

Fig. 3.9 3D view of a runner blade

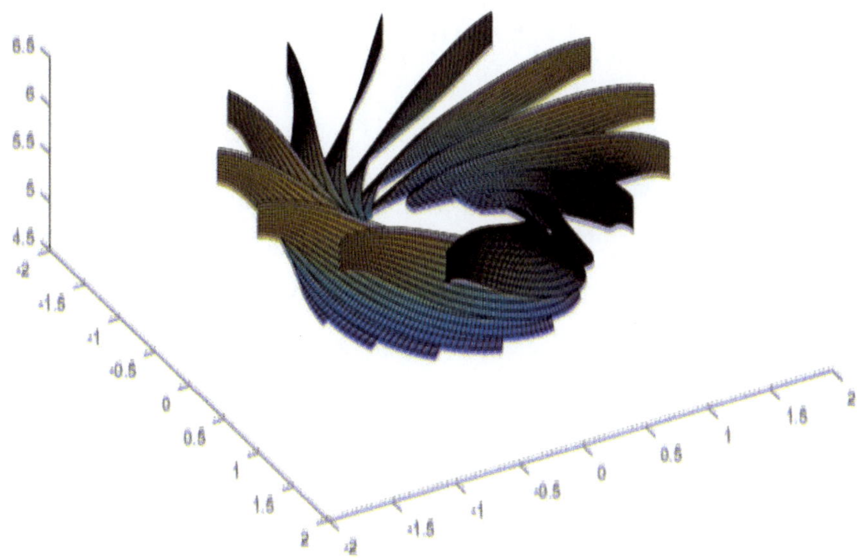

Fig. 3.10 3D view of the runner blades

Chapter 4
Design of the Distributors

4.1 Guide Vanes

The design of guide vanes start with identifying the main dimensions. At this moment, the main dimensions of the runner are already calculated, including D1 and C_{u1}. A vaneless gap is present between the outlet of guide vanes and the inlet of runner. It is possible to calculate the tangential component of absolute velocity at the outlet of the guide vane, C_{ugvo}, by using free vortex theory. Inside the vaneless space between guide vanes and runner, the swirl component of the flow is driven only by the diameter of the turbine; hence, $C_u r = $ constant relation can be followed. Applying this relation in this vaneless space, Eq. (5.1) is obtained.

$$C_{ugvo} = \frac{C_{u1} r_1}{r_{gvo}} \tag{4.1}$$

For high-head Francis turbines, a rule of thumb is that the diameter of the guide vane outlet, D_{gvo}, is 5% more than the diameter of the runner inlet, D_1. Hence, the unknown parameter r_{gvo} in Eq. (4.1) is selected based on the calculated parameter, D_1, to obtain the value of C_{ugvo}. Figure 4.1 shows a 2D view of a turbine illustrating the radius, r_{gvo}.

By using the continuity equation, Eq. (4.2) can be used to obtained the value of C_{mgvo}. As discussed in previous chapters, only the meridional component is responsible for carrying the flow downstream, and the same concept applies for the guide vanes. In this case, B_{gv} is the height of guide vanes, which is equal to B_1 for the case of high-head Francis turbines.

$$C_{mgvo} = \frac{Q}{2\pi B_{gv} r_{gvo}} \tag{4.2}$$

Z. Guo et al., *Design of Francis Turbine and OpenFOAM Methods*,
https://doi.org/10.1007/978-981-99-8381-0_4

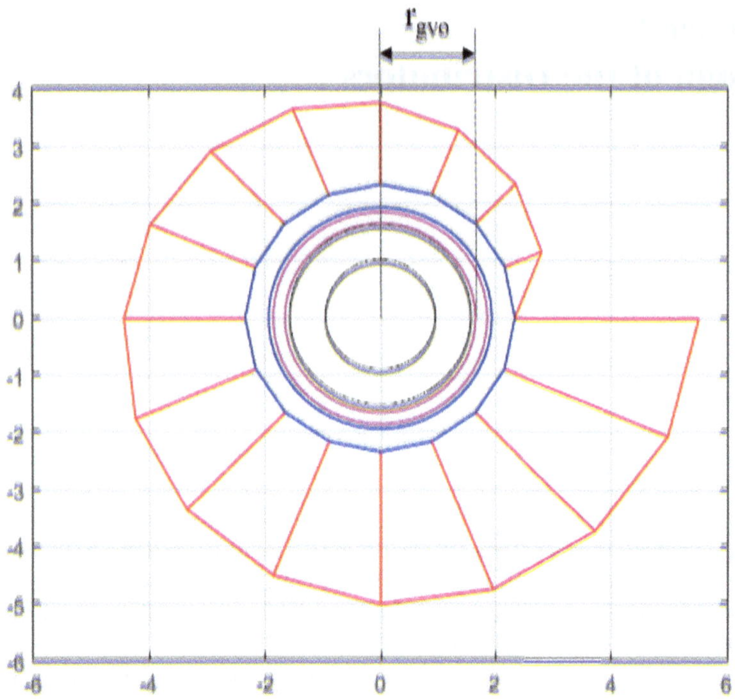

Fig. 4.1 2D view of a turbine indicating r_{gvo}

The flow angle, α_{gvo}, at the outlet of guide vanes is obtained by finding the resultant of the tangential and meridional components, as shown in Eq. (4.3).

$$\tan \alpha_{gvo} = \frac{C_{mgvo}}{C_{ugvo}} \tag{4.3}$$

An empirical relation exists to calculate the initial axis diameter of guide vanes, D_{axf}, as shown in Eq. (4.4). This diameter is necessary to calculate the length of each guide vane, so that there is no gap when the guide vanes are in complete closed position.

$$D_{axf} = D_1(0.29 \cdot \Omega + 1.07) \tag{4.4}$$

In Eq. (4.3), Ω represents the speed number. The number of guide vanes is chosen, which is mostly in the range of 20–24, and Eq. (4.5) is used to calculate the length of each guide vane. In the equation, a coverage factor, k_{cf}, needs to be multiplied, which is generally in the range of 1.10–1.15. This is to have a sufficient overlap between the adjacent guide vanes during the closing position, so that any leakage of water and the full rotation of guide vanes are prevented.

$$L_{gv} = \frac{\pi D_{axf} k_{cf}}{N_{gv}} \tag{4.5}$$

Lastly, by using the cosine theorem, the diameter at the inlet of the guide vanes can be calculated by using Eq. (4.6).

$$D_{gvi} = 2 \times \sqrt{L_{gv}^2 + \left(\frac{D_{gvo}}{2}\right)^2 - 2L_{gv}\left(\frac{D_{gvo}}{2}\right)\cos\left(\frac{\pi}{2} - \alpha_{gvo}\right)} \tag{4.6}$$

Depending upon the profile of guide vanes, it might be necessary to relocate the position of the axis. The most common location is 2/3 of the guide vane length from the leading edge. When changing the axis, it has to be ensured that the new coverage factor is at least 10%. By using the cosine law again, the flow angle at the inlet of guide vane, α_{gvi}, can be calculated.

After calculating the main dimensions of guide vanes, the next task is to give a suitable profile that includes the thickness. Some options of profiling the guide vanes are shown in Fig. 4.2. These profiles are the open-source NACA profiles, which have the maximum thickness of 12% of the chord length. NACA0012 represents a symmetrical profile, whereas others are asymmetrical around the chord. An open-source NACA profile can be resized and repositioned based on the previous calculated values. The final 2D representation after adding the guide vane profile is shown in Fig. 4.3.

Since the profile of the guide vane is same throughout its height, the guide vane surface can be developed based on the known height, its location, and profile. The guide vane surface is shown in Fig. 4.4.

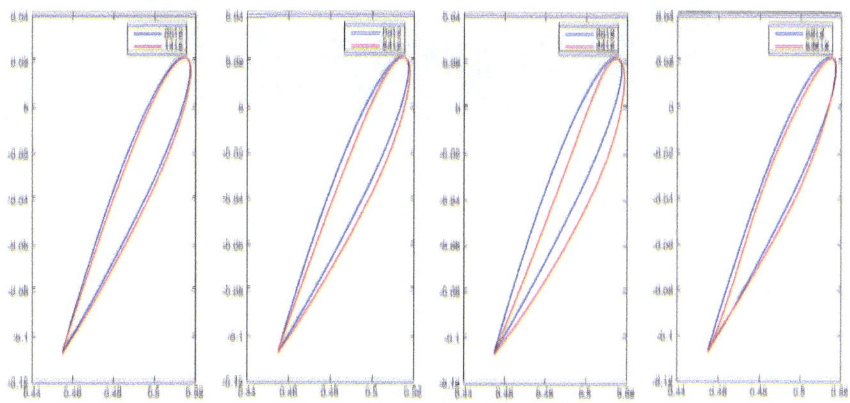

Fig. 4.2 Different guide vane profiles compared with NACA0012

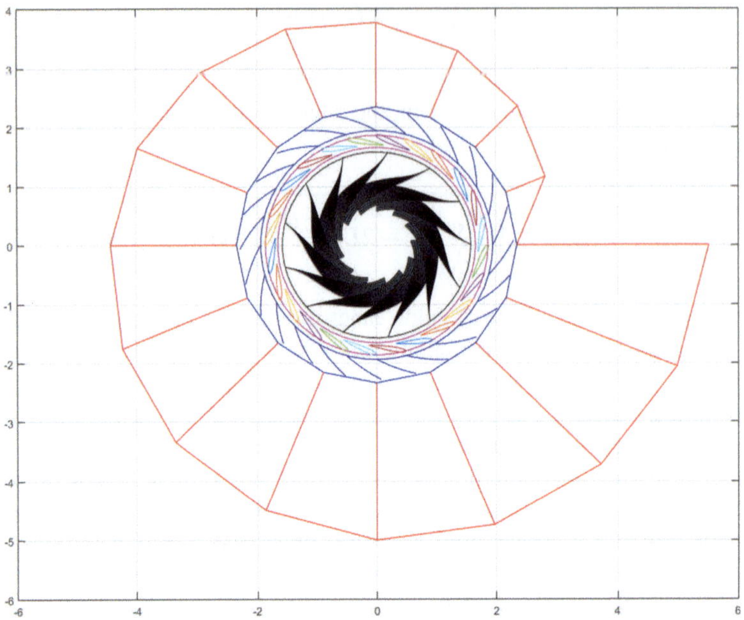

Fig. 4.3 2D representation of the turbine after adding the guide vane profile

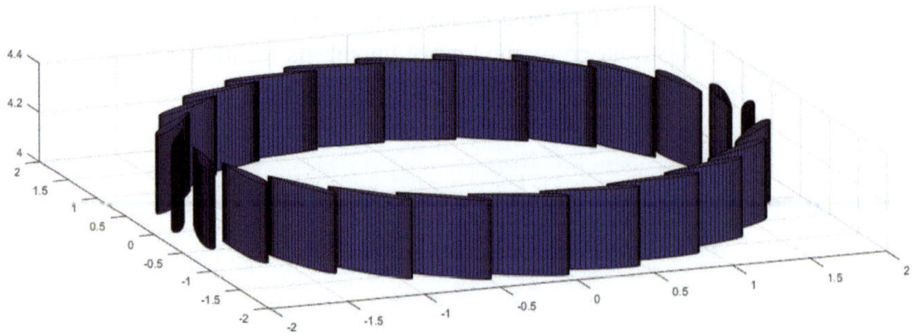

Fig. 4.4 3D representation of the guide vane after adding the profile

4.2 Stay Vanes

Unlike guide vanes, the main purpose of stay vanes is to keep the spiral casing together. Hence, rather than providing an impact hydraulically, stay vanes are mostly designed for adding mechanical strength to the turbine. In order to have a minimum hydraulic effect, the stay vanes are shaped according to the free vortex theory. This means that the flow is guided into the guide vanes along the same path that would be made if the vanes were not present. Based on empirical data, the diameter of the stay vane outlet, D_{svo}, is set to be 2% larger than D_{gvi}. This value can vary between 1 and 5%. Equation (4.1) is the free vortex theory for the vaneless space between stay vanes and guide vanes. Nevertheless, the same equation is also applicable for any radiuses inside the stay vanes.

$$C_{usvo} = \frac{C_{ugvi} r_{gvi}}{r_{svo}} \tag{4.7}$$

Length of the stay vanes is determined based on the total force which they have to resist.

$$L_{sv} = \frac{F_{\max}}{\sigma_{\text{steel}} \cdot t_{sv} \cdot Z_{sv}} \tag{4.8}$$

In Eq. (4.2), σ_{steel} is the property of steel, which is taken to be about 100 MPa. Z_{sv} is the number of stay vanes, which is usually taken same as Z_{gv}. A suitable value of the thickness t_{sv} needs to be chosen to achieve a reasonable L_{sv}. Lower the value of thickness, higher will be the length required to withstand the same force.

$F_{\max}$ in Eq. (4.2) can be calculated based on Eq. (4.3).

$$F_{\max} = P_{\max} \cdot A_{\text{annulus}} \tag{4.9}$$

where

$$A_{\text{annulus}} = \frac{1}{4}\pi \left(D_{svi}^2 - D_{svo}^2\right) \tag{4.10}$$

The area described by Eq. (4.4) is shown in Fig. 4.5. During the calculation of $P_{\max}$ from Eq. (4.3), both hydrostatic and water hammer pressures should be taken into consideration.

$$P_{\max} = P_{\text{head}} + P_{\text{water hammer}} = \rho g H + \frac{2 \cdot \rho \cdot \Delta Q \cdot L_{\text{pipe}}}{T_C \cdot A_{\text{pipe}}} \tag{4.11}$$

In Eq. (4.5), ρ is the density of water, ΔQ is the change of flow rate during shut down, T_C is the closing time of the guide vanes, and L_{pipe} and A_{pipe} are the length and cross-sectional area of the penstock pipe respectively.

4.3 Spiral Casing

Considering the manufacturability of spiral casing, it is divided into a number of sections. The free vortex theory is applied for designing the circular sections. As seen from Fig. 4.6, the following integral relation is used:

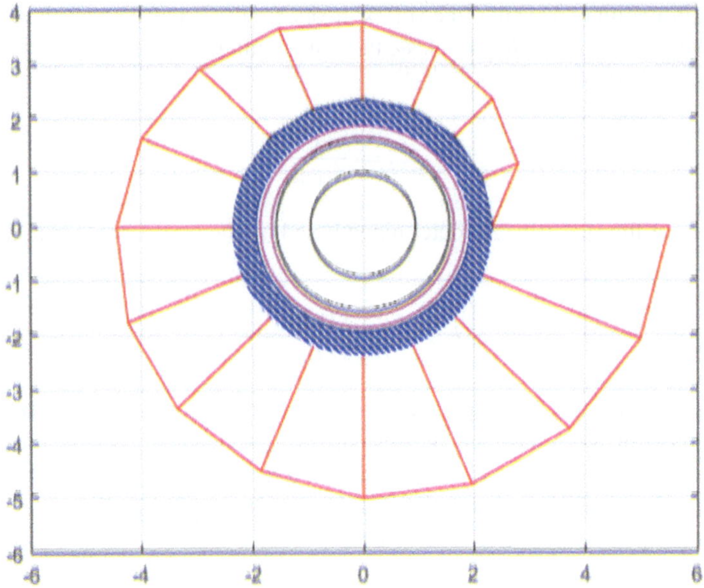

Fig. 4.5 Illustration of A_{annulus} represented by the blue region

$$RC_u = \frac{Q}{2 \cdot r^2 \cdot \int_{\varphi_y}^{\pi} \frac{\sin^2 \varphi}{R_t - r \cdot \cos \varphi} d\varphi} \tag{4.12}$$

$$R_t = R_o + r \cdot \cos(\varphi_o) - k_2 \cdot B \tag{4.13}$$

The parameters in Eqs. (4.12) and (4.13) are explained in Figs. 4.6 and 4.7. The constant k_2 in Eq. (4.13) represents the overlap factor, i.e., if the stay vanes are extended into the spiral casing. Its value is recommended to be 0.1; however, for the sake of simplicity during design, both k_2 and φ_o can be neglected, such that the

Fig. 4.6 Cross section of spiral casing with different dimensions

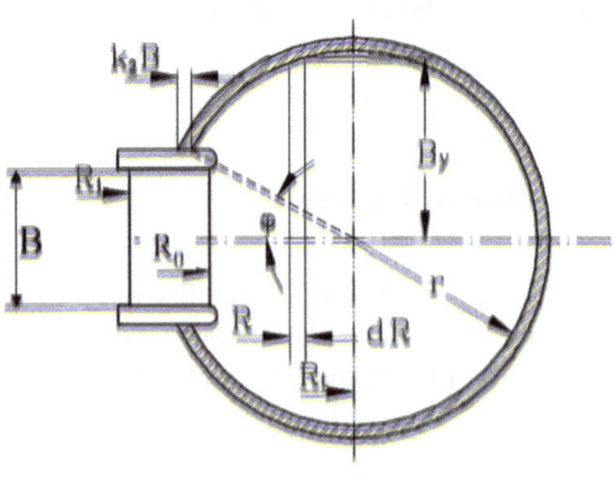

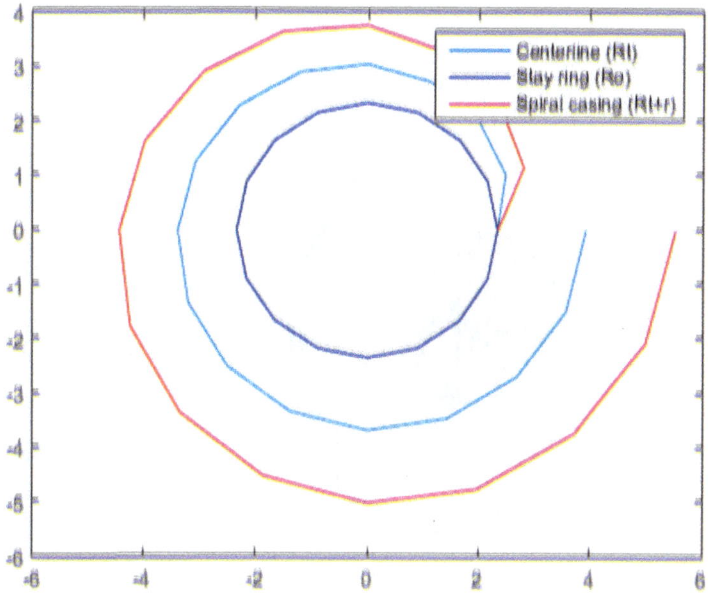

Fig. 4.7 2D of the spiral casing along with the illustration of the terms

equation is reduced to: $R_t = R_o + r$. Here, r is representing the radius of the cross sections, which can be calculated from Eq. (4.6). The term, RC_u, on the left-hand side, remains constant, which can be obtained from the corresponding parameters at the inlet of stay vanes. From the programming point of view, this equation can be solved for r using iterations, such that the equation from both sides is balanced.

4.4 Draft Tube

Draft tubes are conical in shape, but in order to avoid any unfavorable flow patterns, the angle of the cone, γ, i.e., the angle between the centerline and the wall should not be larger than 3°. Low value of the angle indicates that the length of the draft tube needed to recover the kinetic energy at the outlet of the runner to pressure energy is large. The outlet diameter of the draft tube cone can be calculated using Eq. (4.8):

$$D_{\text{cone}\,o} = D_{\text{cone}\,i} + 2 \cdot L_{\text{cone}} . \tan\gamma \tag{4.14}$$

In Eq. (4.8), $D_{\text{cone}\,i} = D_2$, which is the outlet diameter of the runner.

The final assembled components are shown in Fig. 4.8 (without the draft tube).

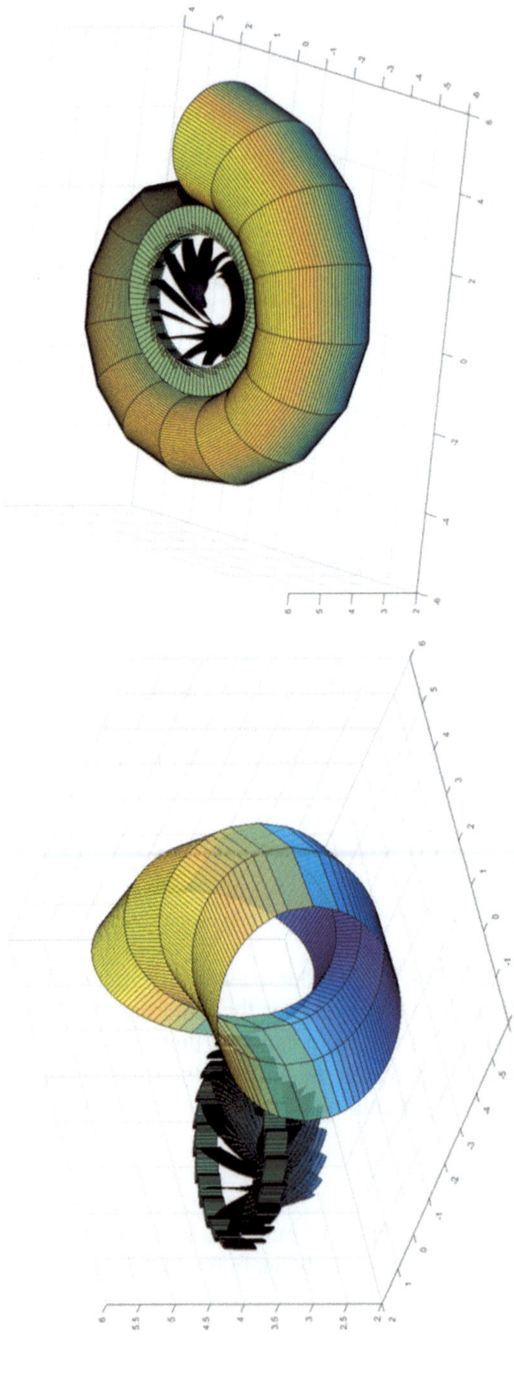

Fig. 4.8 Assembly of the designed components

Chapter 5
Equations, Codes, and Case Study for Single Phase

OpenFOAM is a free, open-source software for computational fluid dynamics (CFD). If one wants to manage it well, the codes must be understood based on governing equations they follow.

5.1 Equations

5.1.1 Governing Equations for Continuous Flow

The steady incompressible governing equations are listed as,

$$\nabla \cdot \mathbf{U} = 0 \tag{5.1}$$

$$\nabla \cdot (\mathbf{UU}) = -\nabla p + \nabla \cdot (\nu \nabla \mathbf{U}) \tag{5.2}$$

where vector **U** represents the absolute velocity, p is pressure divided by density, and ν is kinematic viscosity.

In order to solve the momentum equations, the integral of the quantity over volume is normally converted into integration over the face.

$$\begin{aligned}\int \nabla \cdot (\mathbf{UU})\mathrm{d}V &= \int (\mathbf{UU}) \cdot \mathrm{d}S = \sum \left(\mathbf{U}^{\mathrm{n}}\mathbf{U}^{*}\right)_{f} \cdot \mathbf{S}_f \\ &= \sum \mathrm{F}_f^{\mathrm{n}} \mathbf{U}_f^{*} = \sum \mathrm{F}_f^{\mathrm{n}} \frac{\boldsymbol{U}_N^{*} + \boldsymbol{U}_P^{*}}{2}\end{aligned} \tag{5.3}$$

$$\int \nabla p \mathrm{d}V = \int p \mathrm{d}S = \sum p_f^n \mathbf{S}_f \tag{5.4}$$

Z. Guo et al., *Design of Francis Turbine and OpenFOAM Methods*,
https://doi.org/10.1007/978-981-99-8381-0_5

$$\int \nabla \cdot (\nu \nabla \mathbf{U}) dV = \int (\nu \nabla \mathbf{U}) \cdot dS = \sum \nu (\nabla \mathbf{U}^*)_f \cdot \mathbf{S}_f$$
$$= \sum \nu (\nabla \mathbf{U}^*)_f \cdot \frac{\mathbf{S}_f}{|\mathbf{S}_f|} |\mathbf{S}_f| = \sum \nu \frac{\boldsymbol{U}_N^* - \boldsymbol{U}_P^*}{|\boldsymbol{d}|} |\boldsymbol{S}_f| \tag{5.5}$$

where n means the current time, superscript * means predicted time, and subscript f means value on the face of the mesh. $\mathbf{S}_f$ means face vector, and F is the flux. $\boldsymbol{U}_P^*$ and $\boldsymbol{U}_N^*$ signify the velocity in current and adjacent meshes. $|\boldsymbol{d}|$ is the distance vector between two adjacent meshes.

Take Eqs. (5.3)–(5.5) into Eq. (5.2) to get:

$$\sum F_f^n \frac{\boldsymbol{U}_N^* + \boldsymbol{U}_P^*}{2} = -\sum p_f^n \mathbf{S}_f + \sum \nu \frac{\boldsymbol{U}_N^* - \boldsymbol{U}_P^*}{|\boldsymbol{d}|} |\boldsymbol{S}_f| \tag{5.6}$$

Then,

$$\mathrm{A}_P \boldsymbol{U}_P^* + \sum A_N \boldsymbol{U}_N^* = -\frac{1}{V_P} \sum p_f^n \mathbf{S}_f \tag{5.7}$$

$$\mathrm{A}_P = \frac{1}{V_P} \left(\frac{\sum F_f^n}{2} + \sum \frac{\nu |\boldsymbol{S}_f|}{|\boldsymbol{d}|} \right) \tag{5.8}$$

$$A_N = \frac{1}{V_P} \left(\frac{F_f^n}{2} - \frac{\nu |\boldsymbol{S}_f|}{|\boldsymbol{d}|} \right) \tag{5.9}$$

Equation (5.7) is the momentum predictor, and $\boldsymbol{U}_P^*$ is solved.

According to Eq. (5.7), the following equation can be obtained:

$$A_P \boldsymbol{U}_P^{n+1} + \sum A_N \boldsymbol{U}_N^{n+1} = -\frac{1}{V_P} \sum p_f^{n+1} \mathbf{S}_f \tag{5.10}$$

Then,

$$\boldsymbol{U}_P^{n+1} = -\frac{\sum A_N \boldsymbol{U}_N^{n+1}}{A_P} - \frac{1}{A_P V_P} \sum p_f^{n+1} \mathbf{S}_f \tag{5.11}$$

Set $HbyA_\mathrm{P}^{n+1} = -\frac{\sum A_N \boldsymbol{U}_N^{n+1}}{A_P}$, then,

$$\boldsymbol{U}_P^{n+1} = HbyA_\mathrm{P}^{n+1} - \frac{1}{\mathrm{A_P} V_P} \sum \mathrm{p_f^{n+1}} \mathrm{S_f} \tag{5.12}$$

$$\boldsymbol{U}_{P,f}^{n+1} = HbyA_\mathrm{P,f}^{n+1} - \frac{1}{\mathrm{A_{P,f}}} \left(\frac{1}{V_P} \sum \mathrm{p_f^{n+1}} \mathrm{S_f} \right)_f \tag{5.13}$$

The continuity Eq. (5.1) can be obtained in the discrete form:

$$\sum U_{P,f}^{n+1} \cdot \mathbf{S}_f = 0 \tag{5.14}$$

Combined with Eq. (5.13), it can be written as:

$$\sum\left(HbyA_{\mathrm{P,f}}^{n+1} - \frac{1}{A_{P,f}}\left(\frac{1}{V_P}\sum p_f^{n+1}\mathbf{S}_f\right)_f\right) \cdot \mathbf{S}_f = 0 \tag{5.15}$$

$$\sum HbyA_{\mathrm{P,f}}^{n+1} \cdot \mathbf{S}_f = \sum \frac{1}{A_{P,f}}\left(\frac{1}{V_P}\sum p_f^{n+1}\mathbf{S}_f\right)_f \cdot \mathbf{S}_\mathrm{f} \tag{5.16}$$

$$\nabla \cdot HbyA_{\mathrm{P}}^{n+1} = \nabla \cdot \left(\frac{1}{A_\mathrm{P}}\nabla \mathrm{p}^{\mathrm{n}+1}\right) \tag{5.17}$$

Equation (5.17) is the pressure equation.

5.1.2 SIMPLE Algorithm

The Semi-Implicit Method for Pressure Linked Equations (SIMPLE) algorithm is normally adopted for a steady incompressible, turbulent flow.

By subtracting Eq. (5.7) from Eq. (5.10),

$$A_P U'_P + \sum A_N U'_N = -\frac{1}{V_P}\sum p'_f \mathbf{S}_f \tag{5.18}$$

where $U'_P = U_P^{n+1} - U_P^*$, $U'_N = U_N^{n+1} - U_N^*$, $p'_f = p_f^{n+1} - p_f^n$.

For the SIMPLE algorithm, the second part in the left of Eq. (5.18) is neglected:

$$A_P U'_P = -\frac{1}{V_P}\sum p'_f \mathbf{S}_f \tag{5.19}$$

Combine Eqs. (5.7) and (5.19) to get:

$$\mathrm{A_P} U_P^{**} + \sum A_N U_N^* = -\frac{1}{V_P}\sum \mathrm{p_f^*} S_\mathrm{f} \tag{5.20}$$

where $U_P^{**} = U_P^* + U'_P$, $\mathrm{p_f^*} = \mathrm{p_f^n} + \mathrm{p'_f}$.

Then,

$$U_P^{**} = -\frac{\sum A_N U_N^*}{A_P} - \frac{1}{A_P V_P}\sum \mathrm{p}_f^* \mathbf{S}_f = HbyA_P^* - \frac{1}{A_P V_P}\sum p_f^* \mathbf{S}_f \tag{5.21}$$

$$U^{**}_{P,f} = HbyA^{*}_{P,f} - \frac{1}{A_{P,f}}\left(\frac{1}{V_P}\sum p^{*}_{f}\mathbf{S}_f\right)_f \tag{5.22}$$

The discretized form of the continuity Eq. (5.1) is

$$\sum U^{**}_{P,f} \cdot \mathbf{S}_f = 0 \tag{5.23}$$

Take Eq. (5.22) into Eq. (5.23) to get:

$$\sum\left(\mathrm{Hby}A^{*}_{P,f} - \frac{1}{A_{P,f}}\left(\frac{1}{V_P}\sum p^{*}_{f}\mathbf{S}_f\right)_f\right) \cdot \mathbf{S}_f = 0 \tag{5.24}$$

$$\sum \mathrm{Hby}A^{*}_{P,f} \cdot \mathbf{S}_f = \sum \frac{1}{A_{P,f}}\left(\frac{1}{V_P}\sum p^{*}_{f}\mathbf{S}_f\right)_f \cdot \mathbf{S}_f \tag{5.25}$$

$$\nabla \cdot HbyA^{*}_{\mathrm{P}} = \nabla \cdot \left(\frac{1}{\mathrm{A_P}}\nabla \mathrm{p}^{*}\right) \tag{5.26}$$

Process (as shown in Fig. 5.1):
According to the values in n time step,

1. Calculate $\mathbf{U}^{*}_{p}$ from Eq. (5.7).

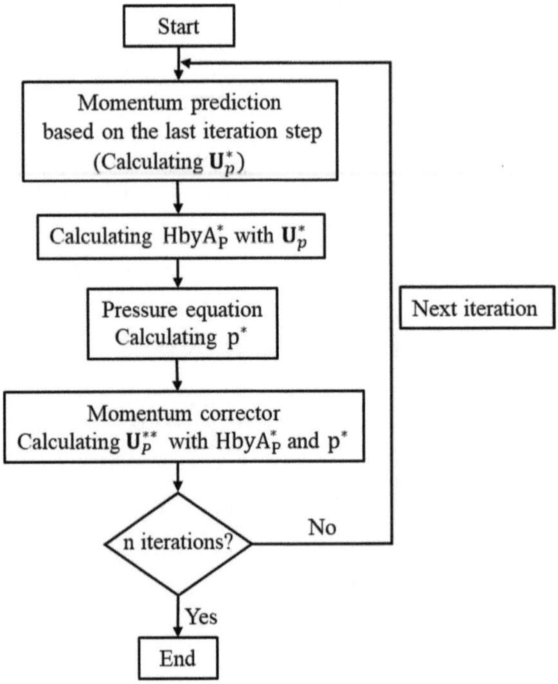

Fig. 5.1 Calculating process of SIMPLE algorithm

2. Calculate HbyA_P^* with $\mathbf{U}_p^*$.
3. Solve Poisson Eq. (5.26) to get p^*.
4. Update $\boldsymbol{U}_P^{**}$ with HbyA_P^* and p^* from Eq. (5.21).
5. Make $\boldsymbol{U}_P^{**}$ as new initial value for next iteration.

5.1.3 SIMPLEC Algorithm

The Semi-Implicit Method for Pressure Linked Equations Consistent (SIMPLEC) algorithm is proposed based on the SIMPLE algorithm.

For SIMPLEC algorithm, the second part in the left of Eq. (5.18) is not neglected but treated as

$$\boldsymbol{U}'_P \approx \frac{\sum A_N \boldsymbol{U}'_N}{\sum \text{A}_\text{N}} \tag{5.27}$$

Take Eq. (5.27) into Eq. (5.18):

$$\text{A}_\text{P}\boldsymbol{U}'_P + \boldsymbol{U}'_P \sum A_N = -\frac{1}{V_P}\sum \text{p}'_\text{f}\boldsymbol{S}_\text{f} \tag{5.28}$$

Then,

$$\boldsymbol{U}'_P = -\frac{1}{\text{A}_\text{P} + \sum A_N}\frac{1}{V_P}\sum p'_f \boldsymbol{S}_f \tag{5.29}$$

Combine Eqs. (5.7) and (5.29):

$$\text{A}_\text{P}\boldsymbol{U}_P^{**} + \sum A_N \boldsymbol{U}_N^* = -\frac{1}{V_P}\sum \text{p}_\text{f}^\text{n}\boldsymbol{S}_\text{f} - \frac{\text{A}_\text{P}}{\text{A}_\text{P} + \sum A_N}\frac{1}{V_P}\sum \text{p}'_\text{f}\boldsymbol{S}_\text{f} \tag{5.30}$$

Then,

$$\begin{aligned}
&\boldsymbol{U}_P^{**} = -\frac{\sum A_N \boldsymbol{U}_N^*}{A_P} - \frac{1}{A_P}\frac{1}{V_P}\sum p_f^n \mathbf{S}_f - \frac{1}{A_P + \sum A_N}\frac{1}{V_P}\\
&\sum p'_f \mathbf{S}_f = -\frac{\sum A_N \boldsymbol{U}_N^*}{A_P} - \frac{1}{A_P}\frac{1}{V_P}\sum p_f^n \mathbf{S}_f + \frac{1}{A_P + \sum A_N}\frac{1}{V_P}\\
&\sum p_f^n \mathbf{S}_f - \frac{1}{A_P + \sum A_N}\frac{1}{V_P}\sum p_f^n \mathbf{S}_f - \frac{1}{A_P + \sum A_N}\frac{1}{V_P}\\
&\sum p'_f \mathbf{S}_f = -\frac{\sum A_N \boldsymbol{U}_N^*}{A_P} - \left(\frac{1}{A_P} - \frac{1}{A_P + \sum A_N}\right)\frac{1}{V_P}\sum p_f^n \mathbf{S}_f - \frac{1}{A_P + \sum A_N}\frac{1}{V_P}\\
&\sum p_f^* \mathbf{S}_f = \text{HbyA}_P^* - \left(\frac{1}{A_P} - \frac{1}{A_P + \sum A_N}\right)\frac{1}{V_P}\sum p_f^n \mathbf{S}_f - \frac{1}{A_P + \sum A_N}\frac{1}{V_P}\sum p_f^* \mathbf{S}_f
\end{aligned} \tag{5.31}$$

Set $HbyA^*_{\text{P,SIMPLEC}} = \text{HbyA}_\text{P}^* - \left(\frac{1}{\text{A}_\text{P}} - \frac{1}{\text{A}_\text{P} + \sum A_N}\right)\frac{1}{V_P}\sum p_f^n \boldsymbol{S}_f$, then,

$$U_P^{**} = HbyA_{P,SIMPLEC}^{*} - \frac{1}{A_P + \sum A_N} \frac{1}{V_P} \sum p_f^* \mathbf{S}_f \tag{5.32}$$

$$U_{P,f}^{**} = HbyA_{P,SIMPLEC,f}^{*} - \frac{1}{A_{P,f} + \sum A_{N,f}} \left(\frac{1}{V_P} \sum p_f^* \mathbf{S}_f \right)_f \tag{5.33}$$

Combining Eq. (5.23), it becomes

$$\sum \left(HbyA_{P,SIMPLEC,f}^{*} - \frac{1}{A_{P,f} + \sum A_{N,f}} \left(\frac{1}{V_P} \sum p_f^* \mathbf{S}_f \right)_f \right) \cdot \mathbf{S}_f = 0 \tag{5.34}$$

$$\sum HbyA_{P,SIMPLEC,f}^{*} \cdot \mathbf{S}_f = \sum \frac{1}{A_{P,f} + \sum A_{N,f}} \left(\frac{1}{V_P} \sum p_f^* \mathbf{S}_f \right)_f \cdot \mathbf{S}_f \tag{5.35}$$

$$\nabla \cdot HbyA_{P,SIMPLEC}^{*} = \nabla \cdot \left(\frac{1}{A_P + \sum A_N} \nabla p^* \right) \tag{5.36}$$

Process:

According to the values in n time step,

1. Calculate $\boldsymbol{U}_p^*$ from Eq. (5.7).
2. Calculate $HbyA_{P,SIMPLEC}^{*}$ with $\boldsymbol{U}_p^*$ and p^n.
3. Solve Poisson Eq. (5.36) to get p^*.
4. Update $\boldsymbol{U}_P^{**}$ with $HbyA_{P,SIMPLEC}^{*}$ and p^* from Eq. (5.32).
5. Make $\boldsymbol{U}_P^{**}$ as the new initial value for the next iteration.

For a moving reference frame rotating with angular velocity $\boldsymbol{\Omega}$ in the absolute velocity formulation, the governing equations of fluid flow can be written as follows:

$$\nabla \cdot \mathbf{U} = 0 \tag{5.37}$$

$$\nabla \cdot (\mathbf{UU}) + \boldsymbol{\Omega} \times \mathbf{U} = -\nabla p + \nabla \cdot (\nu \nabla \mathbf{U}) \tag{5.38}$$

where $\boldsymbol{\Omega} \times \mathbf{U}$ represents the effects of the moving frame.

5.2 Case Study

The following example is presented to use simpleFoam (Appendix-I), so that the flow field inside the nozzle of a Pelton turbine is calculated.

The computational domain is shown in Fig. 5.2, which is divided into five parts, including INLET, NOZZLE, SPEAR, WALL, and OUTLET.

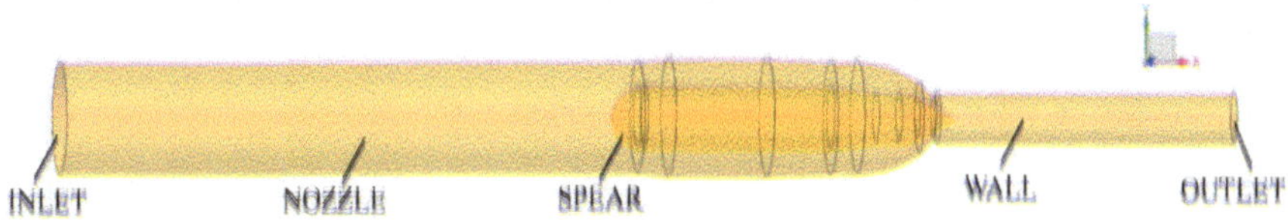

Fig. 5.2 Computational model

5.2.1 *Files Settings*

Three files need to be changed, of which names are 0, constant, and system, separately.

5.2.1.1 File 0

In this file, the initial and boundary conditions are given. Parameters including k, nut, omega, p, and U are set here. Ref. to Fig. 5.3.

File k (turbulent kinetic energy), the initial value of 0.0262 is calculated based on the inlet velocity and length. The calculation formulas are listed as follows [1].

For internal flows, the turbulence intensity I is defined as,

$$I = 0.16(\mathrm{Re})^{-\frac{1}{8}} \tag{5.39}$$

where Re is the Reynolds number.

The turbulent kinetic energy k is defined as,

$$k = \frac{3}{2}(\mathrm{UI})^2 \tag{5.40}$$

where U is the mean flow velocity.

The boundary type of INLET is set as fixedValue, the value of which is set as the same as k. The boundary type of OUTLET is set as zeroGradient. The boundary types of NOZZLE, SPEAR, and WALL are set as kqRWallFunction with the value of uniform 0.File nut, and the initial value is set as uniform 0. The boundary types of INLET and OUTLET are set as calculated with the value of uniform. The boundary types of NOZZLE, SPEAR, and WALL are set as nutkWallFunction with the value of uniform 0.

In the file omega (specific dissipation rate), the initial value is set as 39.33 based on the inlet velocity and length. The calculation formula is shown in Eq. (5.41).

$$\omega = \frac{k^{\frac{1}{2}}}{C_\mu l} \tag{5.41}$$

where ω is the specific dissipation rate. C_μ is an empirical constant specified in the turbulence model. For k = ω turbulence models, this value is 0.09. l is the turbulence

```
kInlet          0.0262;

dimensions      [0 2 -2 0 0 0 0];

internalField   uniform $kInlet;

boundaryField
{
    INLET
    {
        type            fixedValue;
        value           uniform $kInlet;
    }

    OUTLET
    {
        type            zeroGradient;

    }

    NOZZLE
    {
        type            kqRWallFunction;
        value           uniform 0;
    }

    SPEAR
    {
        type            kqRWallFunction;
        value           uniform 0;
    }
    WALL
    {
        type            kqRWallFunction;
        value           uniform 0;
    }
}
```

(a) k

```
dimensions      [0 2 -1 0 0 0 0];

internalField   uniform 0;

boundaryField
{
    INLET
    {
        type            calculated;
        value           uniform 0;
    }

    OUTLET
    {
        type            calculated;
        value           uniform 0;
    }

    NOZZLE
    {
        type            nutkWallFunction;
        value           uniform 0;
    }

    SPEAR
    {
        type            nutkWallFunction;
        value           uniform 0;
    }
    WALL
    {
        type            nutkWallFunction;
        value           uniform 0;
    }

}
```

(b) nut

Fig. 5.3 File 0 settings

```
omegaInlet        39.33;

dimensions        [0 0 -1 0 0 0 0];

internalField     uniform $omegaInlet;

boundaryField
{
    INLET
    {
        type            fixedValue;
        value           uniform $omegaInlet;
    }

   OUTLET
    {
        type            zeroGradient;

    }

    NOZZLE
    {
        type            omegaWallFunction;
        value           uniform 0;
    }

    SPEAR
    {
        type            omegaWallFunction;
         value           uniform 0;
    }
      WALL
    {
        type            omegaWallFunction;
         value           uniform 0;
    }

}
```

(c) omega

```
dimensions        [0 2 -2 0 0 0 0];

internalField     uniform 0;

boundaryField
{
    INLET
    {
        type            zeroGradient;
    }

    OUTLET
    {
        type            totalPressure;
        p0              uniform 0;
    }

    NOZZLE
    {
        type            fixedFluxPressure;
        value           uniform 0;
    }

   SPEAR
    {
        type            fixedFluxPressure;
        value           uniform 0;
    }
     WALL
    {
        type            fixedFluxPressure;
        value           uniform 0;
    }

}
```

(d) p

Fig. 5.3 (continued)

```
Uinlet          (4.20 0 0);
dimensions      [0 1 -1 0 0 0 0];
internalField   uniform $Uinlet;
boundaryField
{
    INLET
    {
        type            fixedValue;
        value           uniform $Uinlet;
    }
    OUTLET
    {
        type            zeroGradient;
    }
    NOZZLE
    {
        type            noSlip;
    }
    SPEAR
    {
        type            noSlip;
    }
    WALL
    {
        type            noSlip;
    }
}
```

(e) U

Fig. 5.3 (continued)

length scale. The value of l is calculated by Eq. (5.42).

$$l = \frac{0.07L}{C_\mu^{\frac{3}{4}}} \tag{5.42}$$

where L is the inlet diameter for this case.

The boundary type of INLET is set as fixedValue, the value of which is set as the same as omega. The boundary type of OUTLET is set as zeroGradient. The boundary types of NOZZLE, SPEAR, and WALL are set as omegaWallFunction with the value of uniform 0.

File p, the initial value is set as uniform 0. The boundary type of INLET is set as zeroGradient. The boundary type of OUTLET is set as totalPressure, where p0 is set as uniform 0. The boundary types of NOZZLE, SPEAR, and WALL are set as fixFluxPressure with the value of uniform 0.

File U, the direction of the inlet is x direction with the value of 4.2 m/s (according to the real situation), and the initial value for the internal field is also set as 4.2 m/s. The boundary type of INLET is set as fixedValue, the value of which is set as the same as U. The boundary type of OUTLET is set as zeroGradient. The boundary types of NOZZLE, SPEAR, and WALL are set as noSlip.

5.2.1.2 Constant Folder

This folder includes 3 files namely polyMesh, momentumTransport, and transportProperties.

(a) polyMesh

(b) momentumTransport

(c) transportProperties

Fig. 5.4 Constant settings

In momentumTransport, the simulation type is RAS, where turbulent mode is SST $k - \omega$.

In transportProperties, the fluid is set as Newtonian, where the kinematic viscosity nu is set as 1.004e−06.

More information is shown in Fig. 5.4.

5.2.1.3 System Folder

The folder includes controlDict, decomposePar, fvSchemes, fvSolution, etc.

In controlDict, simpleFoam is chosen, and the iteration number is set as 15,000.

In decomposePar, the scotch method is chosen, and the value can be given according to the computer's ability and mesh number.

Settings of fvSchemes and fvSolution are shown in Fig. 5.4c, d.

More information is shown in Fig. 5.5.

5.2.2 Results

5.2.2.1 Residual Curve

Notice two aspects, one is to make sure that the error is lower than the given value (here is 1e-6), and the other one is to make sure that the curves are stable with increasing iteration numbers (Fig. 5.6).

```
application     simpleFoam;

startFrom       latestTime;

stopAt          endTime;

endTime         15000;

deltaT          1;

writeControl    timeStep;

writeInterval   3000;

purgeWrite      0;

writeFormat     ascii;

writePrecision   8;

writeCompression off;

timeFormat      general;

timePrecision   6;

runTimeModifiable true;
functions
{
#includeFunc residuals;
}
```

(a) controlDict

```
numberOfSubdomains 16;

method          scotch;
```

(b) decomposePar

```
ddtSchemes
{
    default         steadyState;
}

gradSchemes
{
    default         Gauss linear;

    limited         cellLimited Gauss linear 1;
    grad(U)         $limited;
    grad(k)         $limited;
    grad(omega)     $limited;
}

divSchemes
{
    default         none;

    div(phi,U)      bounded Gauss linearUpwind unlimited;

    turbulence      bounded Gauss linearUpwind limited;
    div(phi,k)      $turbulence;
    div(phi,omega)  $turbulence;
    div(phi,epsilon) $turbulence;
    div((nuEff*dev(T(grad(U))))) Gauss linear;
    div((nuEff*dev2(T(grad(U))))) Gauss linear;
}

laplacianSchemes
{
    default         Gauss linear corrected;
}

interpolationSchemes
{
    default         linear;
}
```

```
snGradSchemes
{
    default         corrected;
}

wallDist
{
    method meshWave;
}
```

(c) fvSchemes

Fig. 5.5 System settings

```
solvers
{
    p
    {
        solver          GAMG;
        smoother        GaussSeidel;
        tolerance       1e-15;
        relTol          0.001;
    }

    "(U|k|omega|epsilon)"
    {
        solver          smoothSolver;
        smoother        symGaussSeidel;
        tolerance       1e-15;
        relTol          0.001;
    }
}

SIMPLE
{
    nNonOrthogonalCorrectors 0;
    consistent      yes;
    momentumPredictor           yes;

    residualControl
    {
        U                       1e-8;
        p                       1e-6;
        "(k|epsilon|omega)"  1e-6;
    }
}

relaxationFactors
{
    fields
    {
        p               0.5;
    }
    equations
    {
        U                   0.7;
        "(k|omega|epsilon)" 0.7;
    }
}
```

(d) fvSolution

Fig. 5.5 (continued)

5.2.2.2 Velocity Contour

See Fig. 5.7.

5.2.2.3 Pressure Contour

See Fig. 5.8.

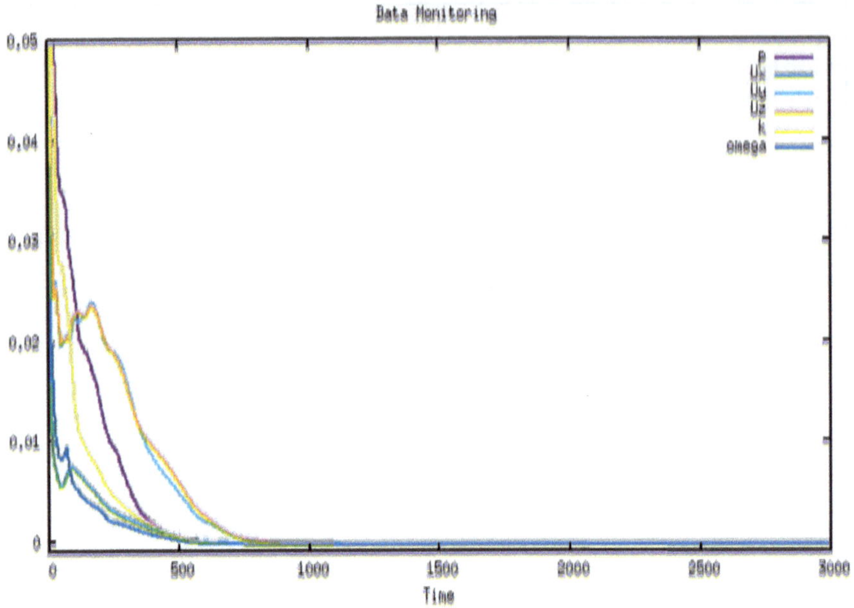

Fig. 5.6 Residual curve

Fig. 5.7 Velocity distribution in the middle slice of the model

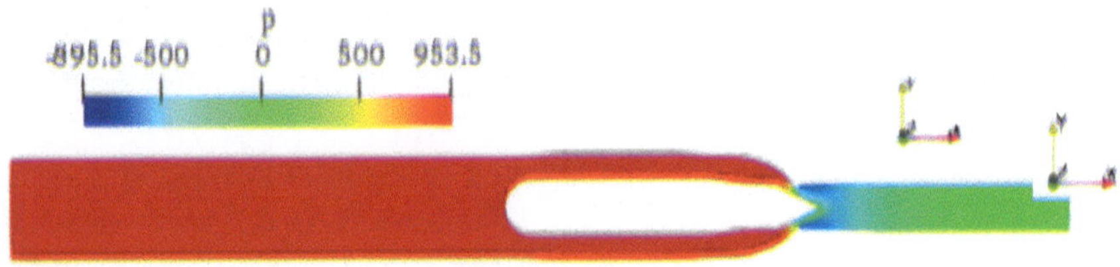

Fig. 5.8 Pressure distribution in the middle slice of the model

Reference

1. ANSYS Fluent, *Fluent 17.2 User's Guide.* ANSYS Inc. (2017)

Chapter 6
Equations, Codes, and Case Study for Three Phases

The version of v2206 in OpenFOAM is installed because the solid–liquid-gas three-phase solver named MPPICInterFoam, which is the coupled transport of a single kinematic particle cloud including the effect of the volume fraction of particles on the continuous phase, is adopted. This solver is developed by both volume of fluid (VOF) method and multiphase particle-in-cell (MPPIC) method, and it is based on the Eulerian–Lagrangian scheme. The solid–liquid-gas three-phase system is controlled by the N-S equation, the motion equation of each particle, and the phase equation of the VOF model.

6.1 Equations

6.1.1 Governing Equations for Continuous Flow

The governing equations are listed as:

$$\frac{\partial \alpha_c \rho}{\partial t} + \nabla \cdot \alpha_c \rho \boldsymbol{U} = 0 \tag{6.1}$$

$$\begin{aligned} &\frac{\partial \alpha_c \rho \boldsymbol{U}}{\partial t} + \nabla \cdot (\alpha_c \rho \boldsymbol{U} \boldsymbol{U}) - \nabla \cdot \alpha_c \boldsymbol{\tau} \\ &= -\alpha_c \nabla p + \alpha_c \rho \boldsymbol{g} + \boldsymbol{F}_s - \boldsymbol{F}_{f-p} \end{aligned} \tag{6.2}$$

where α_c is the volume fraction of the continuous phase in each cell, and the volume fraction of the solid phase is $1 - \alpha_c$. The continuous phase contains the liquid phase (the volume fraction of the liquid phase to the continuous phase is α_1) and gas phase (the volume fraction of the gas phase to the continuous phase $\alpha_2 = 1 - \alpha_1$). ρ is the density of fluid, $\rho = \alpha_1 \rho_1 + \alpha_2 \rho_2$, where the suffixes 1 and 2 denote the liquid and gas phases, respectively. $\boldsymbol{U}$ represents the absolute velocity. p is the pressure

Z. Guo et al., *Design of Francis Turbine and OpenFOAM Methods*,
https://doi.org/10.1007/978-981-99-8381-0_6

divided by density. $\boldsymbol{\tau}$ is shear stress. $\boldsymbol{g}$ is the gravity acceleration. $\boldsymbol{F}_s - \boldsymbol{F}_{f-p}$ is the source term, where the momentum equation considers both surface force and momentum exchange with particles. $\boldsymbol{F}_s$ is the surface force, $\boldsymbol{F}_s = \sigma\kappa\nabla\alpha_1$, where σ is the surface tension and κ is the mean curvature of the free surface. $\boldsymbol{F}_{f-p}$ is the momentum transfer with particles.

In an incompressible flow with ideal conditions, Eq. (6.1) can be simplified as:

$$\frac{\partial\alpha_c}{\partial t} + \nabla\cdot\alpha_c\boldsymbol{U} = 0 \tag{6.3}$$

A modified pressure p_{rgh} is given as:

$$p_{rgh} = p - \rho\boldsymbol{g}\cdot\boldsymbol{h} \tag{6.4}$$

Then,

$$\nabla p_{rgh} = \nabla p - \boldsymbol{g}\cdot\boldsymbol{h}\nabla\rho - \rho\boldsymbol{g} \tag{6.5}$$

The momentum equation is changed as:

$$\begin{aligned} &\frac{\partial\alpha_c\rho\boldsymbol{U}}{\partial t} + \nabla\cdot(\alpha_c\rho\boldsymbol{U}\boldsymbol{U}) - \nabla\cdot\alpha_c\boldsymbol{\tau} \\ &= -\alpha_c\nabla p_{rgh} - \alpha_c\boldsymbol{g}\cdot\boldsymbol{h}\nabla\rho + \sigma\kappa\nabla\alpha_1 - \boldsymbol{F}_{f-p} \end{aligned} \tag{6.6}$$

The calculation of fluid governing equations includes both discretization and solution. Discretization includes three parts: spatial discretization, time discretization, and equation discretization.

The discretized form of momentum equation is obtained:

$$\begin{aligned} a_p\boldsymbol{U}_p^* = &-\sum a_N\boldsymbol{U}_N^* + s_p^t + \nabla\cdot\alpha_c\boldsymbol{\tau}_p^t \\ &- \alpha_c\nabla p_{rgh,p}^t - \alpha_c\boldsymbol{g}\cdot\boldsymbol{h}\nabla\rho_p^t + \sigma\kappa\nabla\alpha_{1,p}^t - \boldsymbol{F}_{f-p,p}^t \end{aligned} \tag{6.7}$$

where the superscripts $*$ and t denote the prediction step and the current time step, respectively.

Equation (6.7) should be reached in the final iteration,

$$\begin{aligned} a_p\boldsymbol{U}_p^{t+\Delta t} &= -\sum a_N\boldsymbol{U}_N^{t+\Delta t} + s_p^t + \nabla\cdot\alpha_c\boldsymbol{\tau}_p^{t+\Delta t} \\ -\alpha_c\nabla p_{rgh,p}^{t+\Delta t} &- \alpha_c\boldsymbol{g}\cdot\boldsymbol{h}\nabla\rho_p^{t+\Delta t} + \sigma\kappa\nabla\alpha_{1,p}^{t+\Delta t} - \boldsymbol{F}_{f-p,p}^{t+\Delta t} \end{aligned} \tag{6.8}$$

Then,

$$\begin{aligned} \boldsymbol{U}_p^{t+\Delta t} &= H_p^{t+\Delta t} \\ &- (a_p)^{-1}\left(\alpha_c\nabla p_{rgh,p}^{t+\Delta t} + \alpha_c\boldsymbol{g}\cdot\boldsymbol{h}\nabla\rho_p^{t+\Delta t} - \sigma\kappa\nabla\alpha_{1,p}^{t+\Delta t} + \boldsymbol{F}_{f-p,p}^{t+\Delta t}\right) \end{aligned} \tag{6.9}$$

where $H_p^{t+\Delta t} = (a_p)^{-1}\left(-\sum a_N U_N^{t+\Delta t} + s_p^t + \nabla \cdot \alpha_c \tau_p^{t+\Delta t}\right)$ and the superscript $t + \Delta t$ denotes the next time step.

The discretized form of Eq. (6.3) is:

$$\frac{\partial \alpha_c}{\partial t} + \sum \alpha_c U_{P,f}^{t+\Delta t} \cdot S_f = 0 \tag{6.10}$$

The discretized form of Eq. (6.3) is:

$$U_{p,f}^{t+\Delta t} = H_{p,f}^{t+\Delta t} - \left(a_{p,f}\right)^{-1}\begin{pmatrix} \alpha_c \nabla_f^{\perp} p_{rgh,p}^{t+\Delta t} + \alpha_c g \cdot h \nabla_f^{\perp} \rho_p^{t+\Delta t} \\ -\sigma\kappa \nabla_f^{\perp} \alpha_{1,p}^{t+\Delta t} + F_{f-p,p,f}^{t+\Delta t} \end{pmatrix} \tag{6.11}$$

Based on Rhie–Chow interpolation, Eq. (6.9) can be read as:

$$\nabla \cdot \left(\left(a_p\right)^{-1} \alpha_c^2 \nabla p_{rgh}^{t+\Delta t}\right) = \frac{\partial \alpha_c}{\partial t} + \nabla \cdot \begin{pmatrix} \alpha_c H_p^{t+\Delta t} \\ -\left(a_p\right)^{-1}\begin{pmatrix} \alpha_c^2 g \cdot h \nabla \rho_p^{t+\Delta t} \\ -\alpha_c \sigma \kappa \nabla \alpha_{1,p}^{t+\Delta t} \\ +\alpha_c F_{f-p,p}^{t+\Delta t} \end{pmatrix} \end{pmatrix} \tag{6.12}$$

Since Eq. (6.12) is unknown for the next time step. The PISO algorithm is then adopted, which obtains the velocity and pressure of the next time step by making multiple corrections to the current time step.

Subtracting (6.7) from Eq. (6.8):

$$\begin{aligned} a_p U'_p = &-\sum a_N U'_N + \nabla \cdot \alpha_c \tau'_p - \alpha_c \nabla p'_{rgh,p} \\ &- \alpha_c g \cdot h \nabla \rho'_p + \sigma\kappa \nabla \alpha'_{1,p} - F'_{f-p,p} \end{aligned} \tag{6.13}$$

Neglect $-\sum a_N U'_N$, then,

$$a_p U'_p = \nabla \cdot \alpha_c \tau'_p - \alpha_c \nabla p'_{rgh,p} - \alpha_c g \cdot h \nabla \rho'_p + \sigma\kappa \nabla \alpha'_{1,p} - F'_{f-p,p} \tag{6.14}$$

Combining Eqs. (6.14) and (6.7):

$$\begin{aligned} a_p U_p^{**} = &-\sum a_N U_N^* + s_p^t + \nabla \cdot \alpha_c \tau_p^* \\ &- \alpha_c \nabla p_{rgh,p}^* - \alpha_c g \cdot h \nabla \rho_p^* + \sigma\kappa \nabla \alpha_{1,p}^* - F_{f-p,p}^* \end{aligned} \tag{6.15}$$

$$H_p^* = \left(a_p\right)^{-1}\left(-\sum a_N U_N^* + s_p^t + \nabla \cdot \alpha_c \tau_p^*\right) \tag{6.16}$$

$$U_p^{**} = H_p^* - \left(a_p\right)^{-1}\left(\alpha_c \nabla p_{rgh,p}^* + \alpha_c g \cdot h \nabla \rho_p^* - \sigma\kappa \nabla \alpha_{1,p}^* + F_{f-p,p}^*\right) \tag{6.17}$$

Based on Rhie–Chow interpolation, Eq. (6.17) can be read as:

$$U_{p,f}^{**} = H_{p,f}^{*} - \left(a_{p,f}\right)^{-1}\left(\alpha_c \nabla_f^{\perp} p_{rgh,p}^{*} + \alpha_c \boldsymbol{g} \cdot \boldsymbol{h} \nabla_f^{\perp} \rho_p^{*} - \sigma\kappa\nabla_f^{\perp}\alpha_{1,p}^{*} + \boldsymbol{F}_{f-p,p,f}^{*}\right) \tag{6.18}$$

The Eq. (6.3) is changed as:

$$\frac{\partial \alpha_c}{\partial t} + \sum \alpha_c \boldsymbol{U}_{P,f}^{**} \cdot \boldsymbol{S}_f = 0 \tag{6.19}$$

Taking Eq. (6.18) into Eq. (6.19) yields the pressure equation:

$$\begin{aligned}\nabla \cdot \left(\left(a_p\right)^{-1}\alpha_c^2 \nabla p_{rgh}^{*}\right) = &\frac{\partial \alpha_c}{\partial t} \\ &+ \nabla \cdot \left(\alpha_c H_P^{*} - \left(a_p\right)^{-1}\begin{pmatrix}\alpha_c^2 \boldsymbol{g} \cdot \boldsymbol{h} \nabla \rho_p^{*} \\ -\alpha_c \sigma\kappa\nabla\alpha_{1,p}^{*} + \alpha_c \boldsymbol{F}_{f-p,p}^{*}\end{pmatrix}\right)\end{aligned} \tag{6.20}$$

The computing cycle is shown in Fig. 6.1,

1. Calculate $\boldsymbol{U}_p^{*}$ from Eq. (6.7).
2. Calculate H_P^{*} with $\boldsymbol{U}_p^{*}$.
3. Solve Eq. (6.20) to get p_{rgh}^{*}.
4. Calculate $\boldsymbol{U}_P^{**}$ with H_P^{*} and p_{rgh}^{*} from Eq. (6.17).
5. Update H_P^{*} with $\boldsymbol{U}_P^{**}$ again and correct the p_{rgh}^{*} (cycle for several iteration times).
6. Make the obtained values as the new initial value for the next time step.

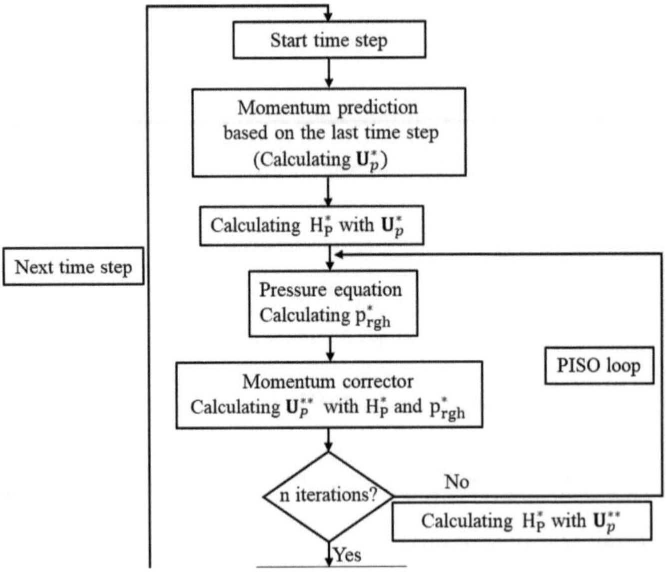

Fig. 6.1 Computing cycle of PISO algorithm

The continuity equations for the liquid and gas phases are especially presented:

$$\frac{\partial \alpha_1}{\partial t} + \nabla \cdot \alpha_1 \boldsymbol{U}_1 = 0 \tag{6.21}$$

$$\frac{\partial \alpha_2}{\partial t} + \nabla \cdot \alpha_2 \boldsymbol{U}_2 = 0 \tag{6.22}$$

$$\alpha_1 + \alpha_2 = 1 \tag{6.23}$$

$$\boldsymbol{U} = \alpha_1 \boldsymbol{U}_1 + \alpha_2 \boldsymbol{U}_2 \tag{6.24}$$

With

$$\begin{aligned} \nabla \cdot \alpha_1 \boldsymbol{U}_1 &= \nabla \cdot \alpha_1 (\boldsymbol{U}_1 + \boldsymbol{U} - \alpha_1 \boldsymbol{U}_1 - \alpha_2 \boldsymbol{U}_2) \\ &= \nabla \cdot \alpha_1 \boldsymbol{U} + \nabla \cdot \alpha_1 (\boldsymbol{U}_1 - \alpha_1 \boldsymbol{U}_1 - (1 - \alpha_1) \boldsymbol{U}_2) \\ &= \nabla \cdot \alpha_1 \boldsymbol{U} + \nabla \cdot (\alpha_1 (1 - \alpha_1))(\boldsymbol{U}_1 - \boldsymbol{U}_2)) \end{aligned} \tag{6.25}$$

Then,

$$\frac{\partial \alpha_1}{\partial t} + \nabla \cdot \alpha_1 \boldsymbol{U} + \nabla \cdot (\alpha_1 (1 - \alpha_1))(\boldsymbol{U}_1 - \boldsymbol{U}_2)) = 0 \tag{6.26}$$

Considering the effect of particle volume:

$$\frac{\partial \alpha_c \alpha_1}{\partial t} + \nabla \cdot \alpha_c \alpha_1 \boldsymbol{U} + \nabla \cdot (\alpha_c \alpha_1 (1 - \alpha_1)(\boldsymbol{U}_1 - \boldsymbol{U}_2)) = 0 \tag{6.27}$$

Equation (6.27) is the alpha equation of the VOF model. The third term is the artificial convection term, which is used to solve the interface ambiguity problem. The relative velocity $\boldsymbol{U}_1 - \boldsymbol{U}_2$ is expressed by $\boldsymbol{U}_r$ as follows:

$$\frac{\partial \alpha_c \alpha_1}{\partial t} + \nabla \cdot \alpha_c \alpha_1 \boldsymbol{U} + \nabla \cdot (\alpha_c \alpha_1 (1 - \alpha_1) \boldsymbol{U}_r) = 0 \tag{6.28}$$

In the actual calculation, $\boldsymbol{U}_r$ is expressed as $\boldsymbol{U}_r = c|\boldsymbol{U}|\frac{\nabla \alpha_1}{|\nabla \alpha_1|}$, c is the compression factor, and the larger c is, the sharper the interface is.

The discretized form of (6.28) is:

$$\frac{\partial \alpha_c \alpha_1}{\partial t} + \sum (\alpha_c \alpha_1 \boldsymbol{U})_f \cdot \boldsymbol{S}_f + \sum (\alpha_c \alpha_1 (1 - \alpha_1) \boldsymbol{U}_r)_f \cdot \boldsymbol{S}_f = 0 \tag{6.29}$$

where $\boldsymbol{S}_f$ is the surface vector.

To ensure that the phase equation is strictly bounded, the MULES algorithm is normally selected.

6.2 Governing Equations for the Particle System

The motion of each particle is controlled by Newton's second law. The following equations can be adopted for particle motion [1, 2]:

$$m\frac{d\boldsymbol{U}_p}{dt} = m\boldsymbol{g} + \sum \boldsymbol{F}_{p-p,1} + \sum \boldsymbol{F}_{p-p,2} + \boldsymbol{F}_{f-p} \tag{6.30}$$

$$I\frac{d\omega}{dt} = \sum M \tag{6.31}$$

where m is the particle mass, $\boldsymbol{U}_p$ is the particle velocity, $\boldsymbol{g}$ is the acceleration of gravity, $\boldsymbol{F}_{p-p,1}$ is the contact force between the particle and the particles, and between the particle and the wall, $\boldsymbol{F}_{p-p,2}$ is the non-contact force between the particle and the particle or other objects, $\boldsymbol{F}_{f-p}$ is the force between the fluid and the particle, including the drag force, fluid stress, etc., I is the rotational inertia, ω is the angular velocity of the particle, and M is the moment acting on this particle by other particles or walls. MPPIC method adopts solids stress terms but not particles collision model to calculate forces between particles.

6.3 Calculation Process and Code Analysis

The calculation process consists of three main steps, as shown in Fig. 6.2. The first step is to solve the particle equation of motion based on the known situation, for calculating the position and velocity of the particle, and this can obtain the particle volume fraction and source term after several times iterations; the second step is to solve the phase equation and update the volume fractions of the liquid and gas phases; thereafter, a PISO loop is performed. After convergence, the fluid-particle interaction force is returned to the particle and the process is repeated.

6.4 Case Study

The following example is presented based on MPPICInterFoam (Appendix-III). The computational domain is shown in Fig. 6.3, which is divided into seven parts, including INLET, NOZZLE, SPEAR, WALL, AIRINLET, AIROUTLET, and OUTLET.

The structured mesh is generated with ICEM, and the number of hexahedrons is 1,580,819. The mesh in the middle slice of the nozzle is shown in Fig. 6.4.

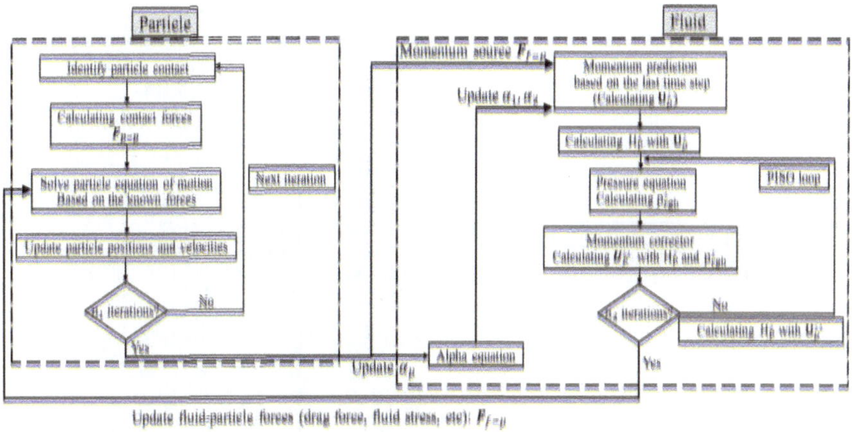

Fig. 6.2 Computing cycle for gas-fluid-solid flow

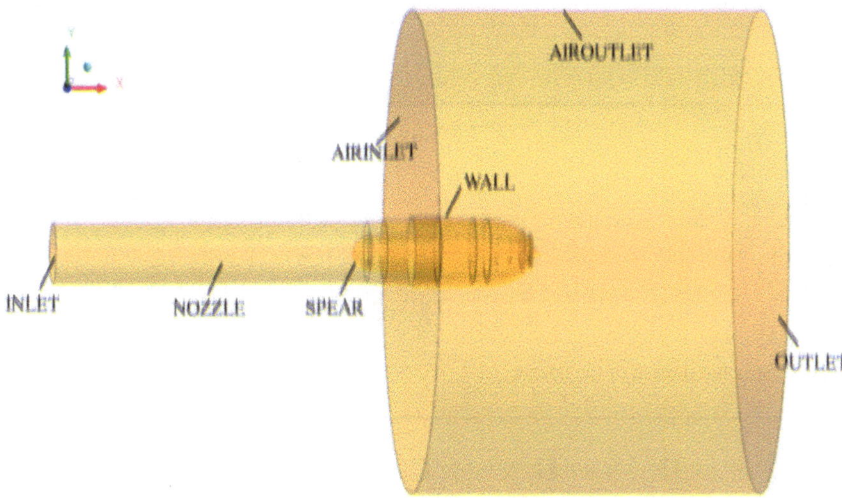

Fig. 6.3 Computational model

6.4.1 Files Settings

6.4.1.1 File 0

In this file, the initial and boundary conditions are given (use the calculation results of InterFoam (step 20,000) as initial conditions), as shown in Fig. 6.5.

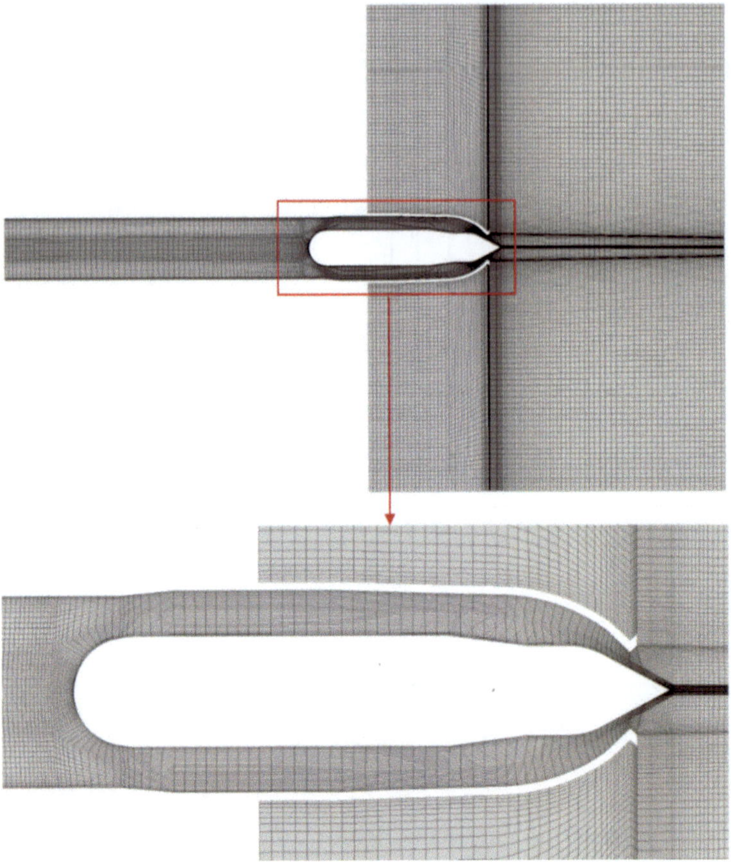

Fig. 6.4 Structured mesh of the nozzle

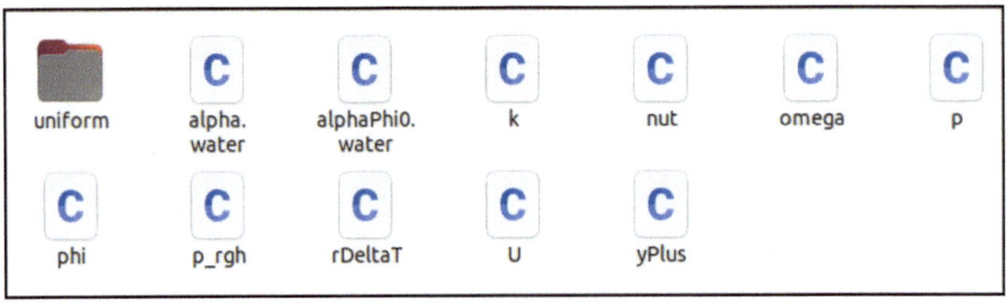

Fig. 6.5 File 0 settings

6.4.1.2 Constant Folder

This location includes files namely polyMesh, g, kinematicCloudProperties, particleTrackDict, transportProperties, and turbulenceProperties.

In g, the gravity acceleration is $9.81\ \mathrm{m/s^2}$.

In kinematicCloudProperties, the particle forces contain drag force and gravity. The parameters related to the injection of particles and erosion wear.

In particleTrackDict, particle trajectories are tracked.

In transportProperties, the water and air are set as Newtonian, where the kinematic viscosity nu is set as 1.003e-06 and 1.48e-05, and the density rho is set as 998.2 and 1.205.

In turbulenceProperties, the simulation type is RAS, where turbulent mode is $k - \omega$.

More information is shown in Fig. 6.6.

6.4.2 System Folder

The folder includes controlDict, decomposePar, fvSchemes, fvSolution, etc.

In controlDict, MPPICInterFoam is chosen, and the iteration number is set as 10,000.

In decomposePar, the scotch method is chosen and 16 is given according to the computer ability and mesh number.

In fvSchemes and fvSolution, the discretization method and solution method are set. For example, the compression factor c is set in "cAlpha" in fvSolution.

More information is shown in Fig. 6.7.

6.4.3 Results

6.4.3.1 Residual Curve

See Fig. 6.8.

6.4.3.2 Volume Fraction Distribution of Continuous Phase

See Fig. 6.9.

6.4.3.3 Velocity Contour

See Fig. 6.10.

(a) polyMesh

```
    dimensions      [0 1 -2 0 0 0 0];
    value           (0 -9.81 0);
```

(b) g

```
subModels
{
    particleForces
    {
        WenYuDrag
        {
            alphac      alphac;
        }
        gravity;
        interface
        {
            C            -10;
            alpha        alpha.water;
        }
    }
```

```
 injectionModels
 {
   model1
    {
           type                 patchInjection;
           parcelBasisType      mass;
           patch                INLET;
           SOI                      20000;
           U0                   (4.20 0 0);
           duration                     1;
           parcelsPerSecond        600000;
           flowRateProfile      constant 1;
           massFlowRate            0.4528;

           sizeDistribution
             {
                 type           RosinRammler;
                 RosinRammlerDistribution
                 {
                     minValue  0.000000632;
                     maxValue  0.001124683;
                     d            0.0002157;
                     n                1.484;
                 }
             }
    }
 }
```

Fig. 6.6 Constant settings

6.4.3.4 Pressure Contour

See Fig. 6.11.

```
cloudFunctions
{
        Erosion
        {
                type                    ParticleErosionOka;
                patches                 (SPEAR NOZZLE);
                 Hv                     1.2;
                 rhoM                   7700;

        }
        Averagevalue
        {
                type                    AverageValue;
                patches                 (SPEAR NOZZLE);
        }

        Track
        {
                type                    particleTracks;
                trackInterval           4;
                maxSamples              100000;
                resetOnWrite            yes;
        }
        patchPost
        {
                type                    patchPostProcessing;
                maxStoredParcels        100000;
                patches                 (SPEAR NOZZLE);
                resetOnWrite            yes;
        }
}
```

(c) kinematicCloudProperties

```
cloud       kinematicCloudTracks;

fields
(
    d
    U
    origId
    nRel
);
```

(d) particleTrackDict

```
phases (water air);

water
{
    transportModel  Newtonian;
    nu              1.003e-06;
    rho             998.2;
}

air
{
    transportModel  Newtonian;
    nu              1.48e-05;
    rho             1.205;
}

sigma           0.0728;
```

(e) transportProperties

```
simulationType      RAS;

RAS
{
    RASModel        kOmega;

    turbulence      on;

    printCoeffs     on;
}
```

(f) turbulenceProperties

Fig. 6.6 (continued)

```
application      MPPICInterFoam;
startFrom        latestTime;
startTime        20000;
stopAt           endTime;
endTime          30000;
deltaT           1;
writeControl     timeStep;
writeInterval    1000;
purgeWrite       0;
writeFormat      binary;
writePrecision    6;
writeCompression off;
timeFormat       general;
timePrecision    6;
runTimeModifiable yes;
adjustTimeStep   yes;
functions
{
#includeFunc residuals;
}
```

(a) controlDict

```
numberOfSubdomains 16;
method          scotch;
```

(b) decomposePar

Fig. 6.7 System settings

6.4.3.5 Erosion Rate Contour

See Fig. 6.12.

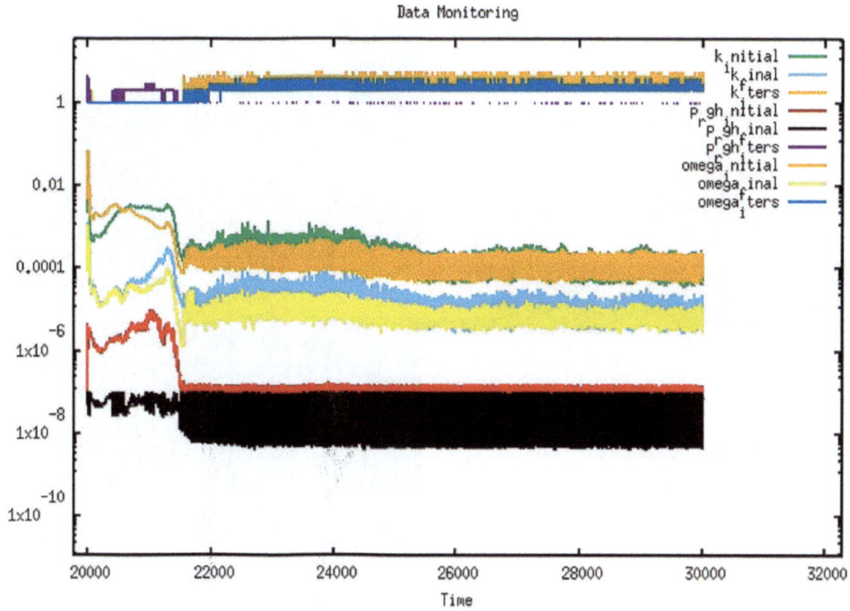

Fig. 6.8 Residual curve

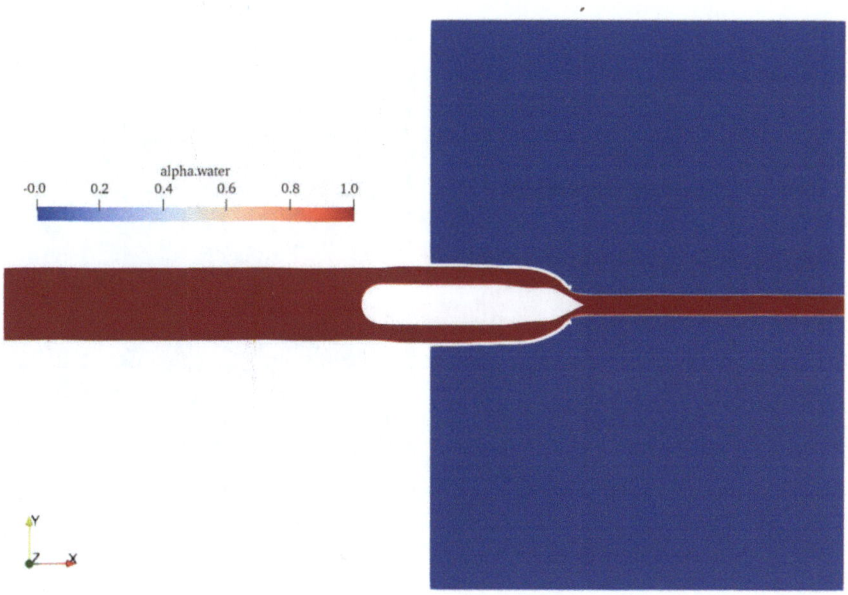

Fig. 6.9 Volume fraction distribution of continuous phase in the middle slice of the model

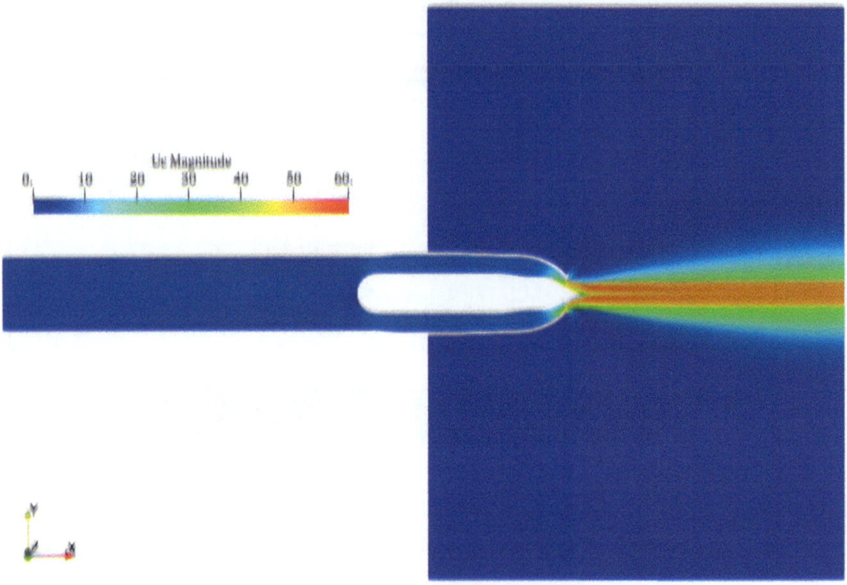

Fig. 6.10 Velocity distribution in the middle slice of the model

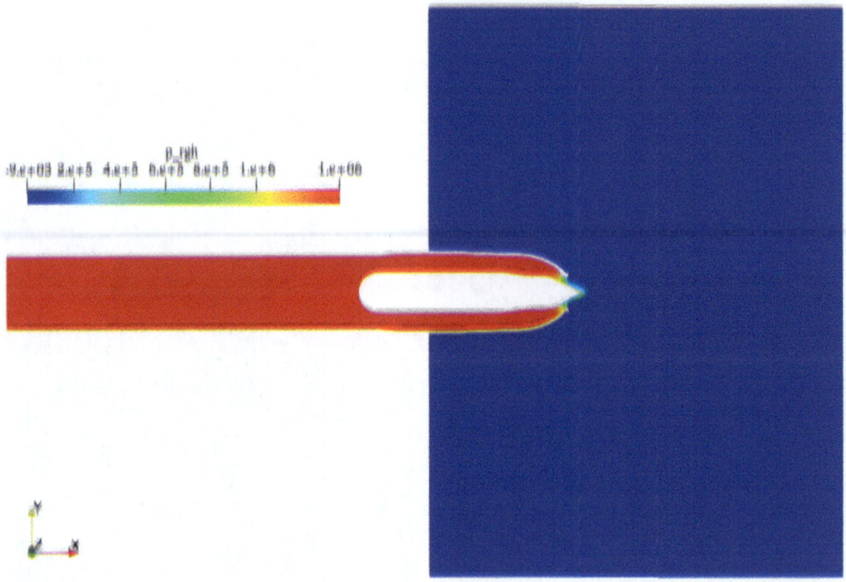

Fig. 6.11 p_{rgh} distribution in the middle slice of the model

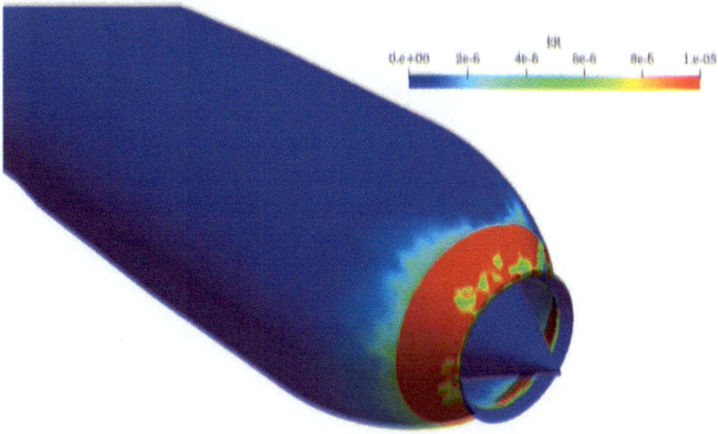

(a) nozzle

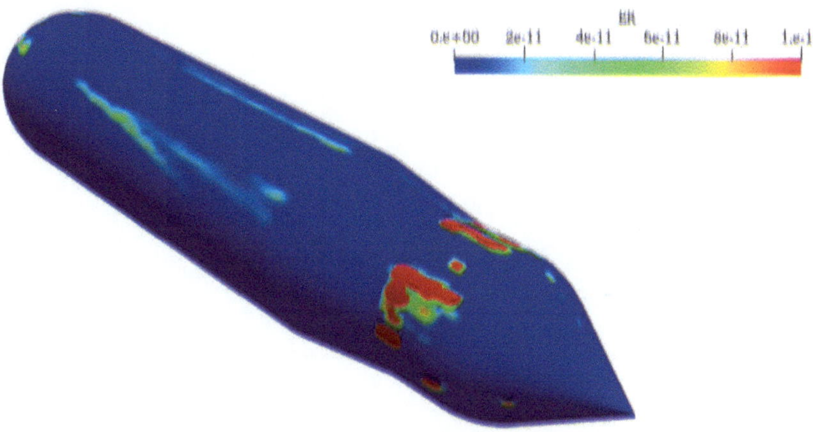

(b) spear

Fig. 6.12 Erosion rate contour of nozzle and spear

References

1. L. Jing, C.Y. Kwok, Y.F. Leung, Y.D. Sobral, Extended CFD-DEM for free-surface flow with multi-size granules. Int. J. Numer. Anal. Meth. Geomech. **40**(1), 62–79 (2016)
2. C. Fernandes, D. Semyonov, L.L. Ferrás, J.M. Nóbrega, Validation of the CFD-DPM solver DPMFoam in OpenFOAM® through analytical, numerical and experimental comparisons. Granular Matt. **20**(4), 64 (2018)

Chapter 7
Erosion Calculation in Francis Turbine

The rotating runner of a Francis turbine is chosen as a reference case for this chapter. The software used for the CFD calculations is OpenFOAM-8 (v8), but other versions can also be chosen. The solver adopted for steady calculation of erosion is a revised one by incorporating discrete phase model (DPM) into SRFSimpleFoam. The steady calculation is conducted in two stages: one is operated with clear water, and the other is operated with both particles and water. The main aim of this chapter is to present a way to calculate erosion and how to make a new solver step by step, where the main settings and codes are also shown.

7.1 Single Phase with Only Clear Water

The runner model is shown in Fig. 7.1, where the number of blades is 17 and the rotational speed is 1000 rpm (rev/min). The model consists of a crown (hub), a band (shroud), runner blades, and inlet and outlet surfaces.

7.1.1 Pre-processing

The procedure for this stage is presented below,

1. File creation.
2. Mesh generation.
3. Physical properties and turbulent model.
4. Boundary and initial conditions.
5. Numerical schemes.
6. Solution and algorithm control.
7. Running the case.

Z. Guo et al., *Design of Francis Turbine and OpenFOAM Methods*,
https://doi.org/10.1007/978-981-99-8381-0_7

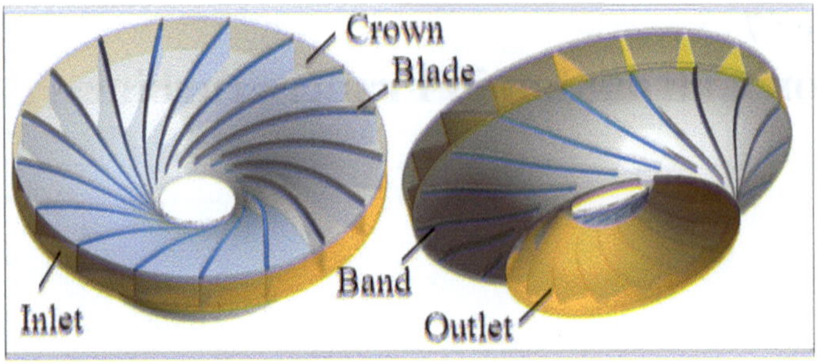

Fig. 7.1 Computational model of Francis turbine

7.1.1.1 File Creation

Firstly, create a file for editing case filescalled runner-clearwater. Secondly, copy "OpenFOAM-8/tutorials/incompressible/ SRFSimpleFoam/mixer" to the current case file. In general practice of OpenFOAM, copying similar cases from tutorials helps to reduce the setting-up time.

7.1.1.2 Mesh Generation

Move the runner.msh that has been created to the case directory of "runner-clearwater". Open the terminal and type the following to make the mesh suitable for OpenFOAM.

```
fluentMeshToFoam runner.msh
```

Check the mesh quality by typing checkMesh, and the output is shown in Fig. 7.2.

7.1.1.3 Physical Properties and Turbulent Model

1. Physical Properties

The physical properties are stored in transportProperties.

Check that if the kinematic viscosity is set correctly by opening the transportProperties dictionary, as shown in Fig. 7.3.

The kinematic viscosity is 1e−06 m^2/s, and one can change the value according to the real situation. The transportModel is Newtonian, and the nu is set as [0 2 − 1 0 0 0 0] (dimensions) 1e−06 (value).

2. Turbulent Model

The turbulent model is stored in momentumTransport.

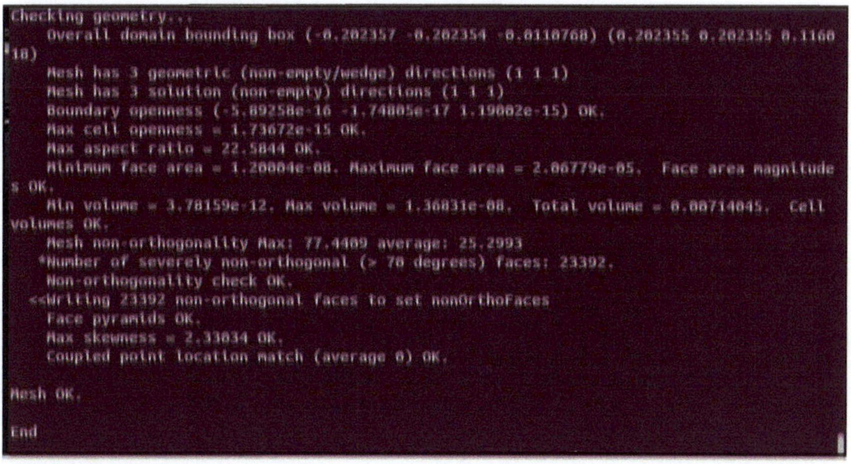

Fig. 7.2 Mesh checked

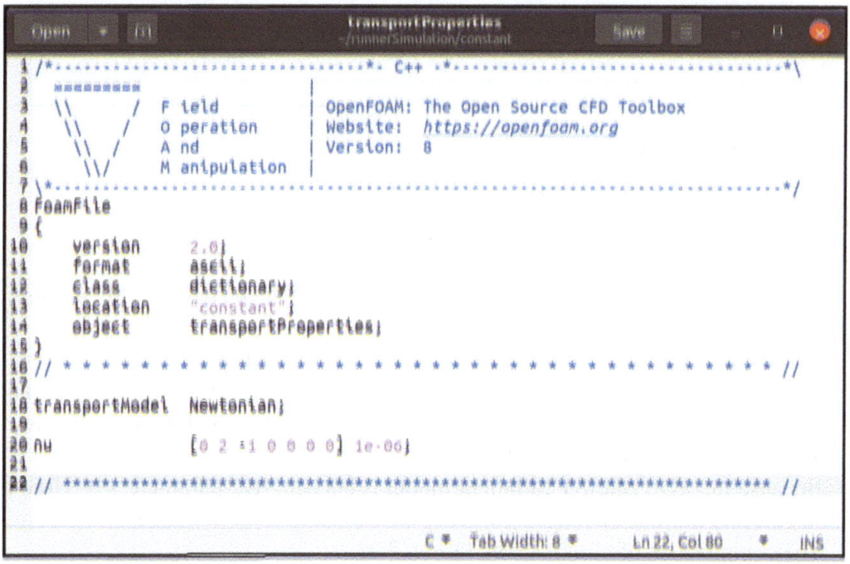

Fig. 7.3 TransportProperties

The kOmegaSST turbulent model is used, and the corresponding momentumTransport file is shown in Fig. 7.4 where the simulationType is RAS (RANS).

3. Moving Frame

The moving frame in the current case is a single rotating frame, which is set in dictionaries SRFProperties. The origin is (0 0 0), the axis is (0 0 − 1), and the rotation speed is 1000 rpm, as shown in Fig. 7.5.

```
/*--------------------------------*- C++ -*----------------------------------*\
  =========                 |
  \\      /  F ield         | OpenFOAM: The Open Source CFD Toolbox
   \\    /   O peration     | Website:  https://openfoam.org
    \\  /    A nd           | Version:  8
     \\/     M anipulation  |
\*---------------------------------------------------------------------------*/
FoamFile
{
    version     2.0;
    format      ascii;
    class       dictionary;
    location    "constant";
    object      momentumTransport;
}
// * * * * * * * * * * * * * * * * * * * * * * * * * * * * * * * * * * * * * //

simulationType RAS;

RAS
{
    model           kOmegaSST;

    turbulence      on;

    printCoeffs     on;
}


// ************************************************************************* //
```

Fig. 7.4 MomentumTransport

7.1.1.4 Boundary and Initial Conditions

Once the mesh generation is complete as shown in Fig. 7.6, the initial fields are stored in a subdirectory named 0.

The boundary setting includes four parts as shown in Fig. 7.7. The number 21 means the number of boundaries, i.e., 17 BLADEs, INLET, OUTLET, HUB, and SHROUD. All of them need to be set with a suitable type. INLET means entrance of fluid. OUTLET means exit of fluid. HUB and SHROUD are wall.

Because the kOmegaSST turbulent model is chosen, five parameters of Urel, p, k, omega, and nut must be set.

The complete set of Urel is given in Fig. 7.8.

The internal field is uniform, described by a single value. The type of INLET is set as cylindricalInletVelocity, where the origin is (0 0 0), the axis is (0 0 1), and axial component (axialVelocity) is constant with the value of 0. The radial component (radialVelocity) is −4.446 m/s, and rotational component (rpm) is 246.656 rpm. The three components of inlet velocity can be calculated based on velocity triangle. The type of OUTLET is set as pressureInletOutletVelocity, and the value is uniform (0 0 0). The type of wall is set as noSlip.

The complete set of p is given in Fig. 7.9.

```
/*--------------------------------*- C++ -*----------------------------------*\
  =========                 |
  \\      /  F ield         | OpenFOAM: The Open Source CFD Toolbox
   \\    /   O peration     | Website:  https://openfoam.org
    \\  /    A nd           | Version:  8
     \\/     M anipulation  |
\*---------------------------------------------------------------------------*/
FoamFile
{
    version     2.0;
    format      ascii;
    class       dictionary;
    location    "constant";
    object      SRFProperties;
}
// * * * * * * * * * * * * * * * * * * * * * * * * * * * * * * * * * * * * * //

SRFModel        rpm;

origin          (0 0 0);
axis            (0 0 -1);

rpmCoeffs
{
    rpm         1000;
}


// ************************************************************************* //

```

Fig. 7.5 SRFProperties

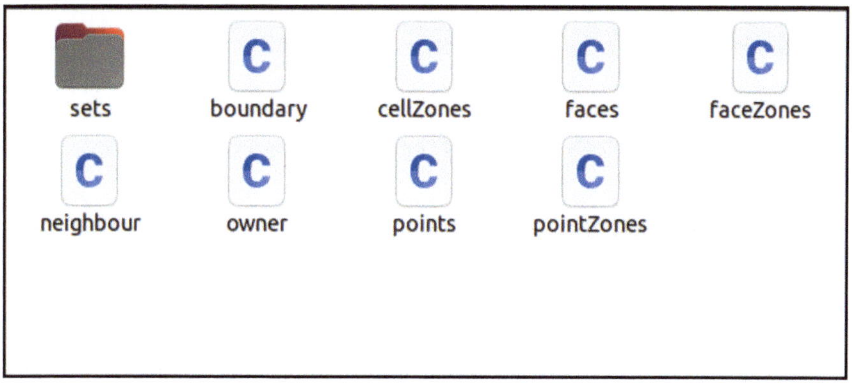

Fig. 7.6 Mesh

The type of INLET is set as zeroGradient (Neumann), and the type of OUTLET is set as fixedValue (Direchlet), where the value is uniform 0. The type of wall is set zeroGradient.

The complete set of k is given in Fig. 7.10.

```
21
(
    INLET
    {
        type            patch;
        nFaces          22542;
        startFace       7678186;
    }
    OUTLET
    {
        type            patch;
        nFaces          22542;
        startFace       7700728;
    }
    SHROUD
    {
        type            wall;
        inGroups        List<word> 1(wall);
        nFaces          76840;
        startFace       7723270;
    }
    HUB
    {
        type            wall;
        inGroups        List<word> 1(wall);
        nFaces          76840;
        startFace       7800110;
    }
```

Fig. 7.7 Boundary type

```
dimensions      [0 1 -1 0 0 0 0];

internalField   uniform (0 0 0);

boundaryField
{
    INLET
    {
        type            cylindricalInletVelocity;
        origin          (0 0 0);
        axis            (0 0 1);
        axialVelocity   constant 0;
        radialVelocity  constant -4.446;
        rpm             constant 246.656;
        value           uniform (0 0 0);
    }
    OUTLET
    {
        type            zeroGradient;
    }
    wall
    {
        type            noSlip;
    }
}
```

Fig. 7.8 Urel

```
dimensions      [0 2 -2 0 0 0 0];

internalField   uniform 0;

boundaryField
{
    INLET
    {
        type            zeroGradient;
    }
    OUTLET
    {
        type            fixedValue;
        value           uniform 0;
    }
    wall
    {
        type            zeroGradient;
    }
}
```

Fig. 7.9 File p

```
dimensions      [0 2 -2 0 0 0 0];

internalField   uniform 0.289;

boundaryField
{
    INLET
    {
        type            fixedValue;
        value           uniform 0.289;
    }
    OUTLET
    {
        type            inletOutlet;
        inletValue      uniform 0.289;
    }
    wall
    {
        type            kqRWallFunction;
        value           uniform 0.289;
    }
}
```

Fig. 7.10 File k

The internal field data are set as uniform 0.289, which is calculated based on inlet velocity and specific length.

The type of INLET is set as fixedValue, and the value is uniform 0.289. The type of OUTLET is set as inletOutlet, and the value is uniform 0.289. The type of wall is set as kqRWallFunction, and the value is uniform 0.289.

The complete set of omega is given in Fig. 7.11.

```
dimensions      [0 0 -1 0 0 0 0];

internalField   uniform 35;

boundaryField
{
    INLET
    {
        type            fixedValue;
        value           uniform 35;
    }
    OUTLET
    {
        type            inletOutlet;
        inletValue      uniform 335;
    }
    wall
    {
        type            omegaWallFunction;
        value           uniform 35;
    }
}
```

Fig. 7.11 Omega

The internal field data of omega are set as uniform 35, which is also calculated based on the inlet velocity and the specific length.

The type of INLET is set as fixedValue, and the value is uniform 35. The type of OUTLET is set as inletOutlet, and the value is uniform 35. The type of wall is set as omegaWallFunction, and the value is uniform 35.

The complete set of nut is given in Fig. 7.12. The internal field data of nut are set as uniform 0.

```
dimensions      [0 2 -1 0 0 0 0];

internalField   uniform 0;

boundaryField
{
    INLET
    {
        type            calculated;
        value           uniform 0;
    }
    OUTLET
    {
        type            calculated;
        value           uniform 0;
    }
    wall
    {
        type            nutkWallFunction;
        value           uniform 0;
    }
}
```

Fig. 7.12 Nut

```
ddtSchemes
{
    default         steadyState;
}

gradSchemes
{
    default         cellLimited Gauss linear 0.5;
    grad(k)            cellLimited Gauss linear 1;
    grad(omega)        cellLimited Gauss linear 1;
}

divSchemes
{
    default             none;
    div(phi,Urel)       bounded Gauss linearUpwind grad(Urel);
    div(phi,k)         bounded Gauss limitedLinear 1;
    div(phi,omega)     bounded Gauss limitedLinear 1;
    div((nuEff*dev2(T(grad(Urel))))) Gauss linear;
}

laplacianSchemes
{
    default         Gauss linear corrected;
}

interpolationSchemes
{
    default         linear;
}

snGradSchemes
{
    default         corrected;
}

wallDist
{
    method meshWave;
}
```

Fig. 7.13 File fvSchemes

7.1.1.5 Numerical Schemes

Numerical schemes must be specified for further calculation, which is subdivided within the fvSchemes dictionary, including ddtSchemes, gradSchemes, divSchemes, laplacianSchemes, interpolationSchemes, snGradSchemes, and wallDist, and the detailed setting is shown in Fig. 7.13.

The ddtSchemes is set for first and second time derivatives if the simulation is unsteady. Otherwise, one can choose steadyState for steady flow.

7.1.1.6 Solution and Algorithm Control

The complete set of fvSolution is given in Fig. 7.14.

```
solvers
{
    "(p|Phi)"
    {

        solver          GAMG;
        tolerance       1e-015;
        relTol          0.05;
        smoother        GaussSeidel;
        nCellsInCoarsestLevel 20;
        maxIter         1000;
        tolerance       1e-05;

    }

    "(Urel|k|omega)"
    {
        solver          smoothSolver;
        smoother        GaussSeidel;
        tolerance       1e-15;
        relTol          0;
        nSweeps         1;
    }
}
SIMPLE
{
    nNonOrthogonalCorrectors    2;
    consistent                  yes;
    momentumPredictor           yes;

    residualControl
    {
        p                       1e-4;
        Urel                    1e-5;
        "(k|omega)"             1e-10;
    }
}

relaxationFactors
{
    fields
    {
        p                       0.5;
    }
    equations
    {
        Urel                    0.3;
        "(k|omega)"             0.3;
    }

cache
{
    grad(Urel);
}

}
```

Fig. 7.14 File fvSolution

```
application     SRFSimpleFoam;
startFrom       startTime;
startTime       0;
stopAt          endTime;
endTime         8000;
deltaT          1;
writeControl    timeStep;
writeInterval   2000;
purgeWrite      0;
writeFormat     ascii;
writePrecision  6;
writeCompression off;
timeFormat      general;
timePrecision   6;
runTimeModifiable true;
```

Fig. 7.15 File controlDict

An efficiency, stable and high-resolution setting of fvSolution is normally based on lots of tests, which includes values of relaxationFactors. One may first learn from tutorials and change the value according to the real situation.

The complete set of controlDict is given in Fig. 7.15. Notice that SRFSimpleFoam is chosen, one can change the values of endTime and writeInterval.

7.1.1.7 Running the Case

1. Parallel Setting

Parallel calculation is widely adopted to save time since the grid number is large. In OpenFOAM, set the number of cores that are needed. Then, 8 is chosen in this case, as shown in Fig. 7.16.

The files (processor 0, processor 1, …, processor 7) can be found in Fig. 7.17.

2. Monitor Setting

Search files of patchAverage and residuals in OpenFOAM-version, and copy the two files to the case directory of system. Open patchAverage, change the name to OUTLET, and add Urel in the fields. Open residuals, and add p and Urel in fields.

Open controlDict, add the codes in the end of the file as follows:

```
FoamFile
{
    version     2.0;
    format      ascii;
    class       dictionary;
    note        "mesh decomposition control dictionary";
    object      decomposeParDict;
}
// * * * * * * * * * * * * * * * * * * * * * * * * * * * * * * * * * * * * * * //

numberOfSubdomains  8;

method scotch;
```

Fig. 7.16 DecomposeParDict

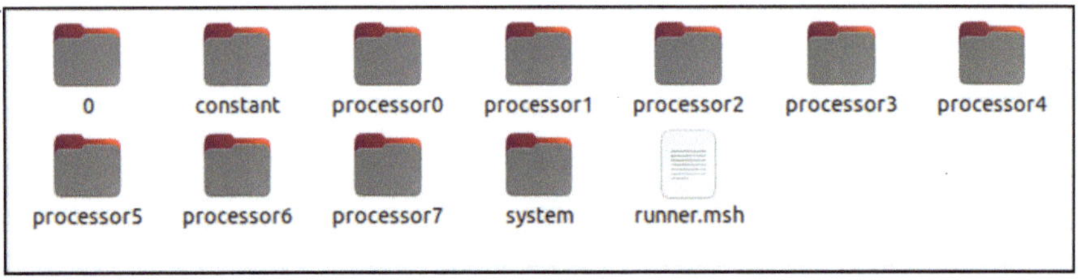

Fig. 7.17 Subfiles for parallel setting

```
functions
{
    #includeFunc residuals;
    #includeFunc patchAverage;
}
```

.

3. Case Running and Residuals Monitoring

Type the following code in the terminal:

```
mpirun -np 8 SRFSimpleFoam -parallel
foamMonitor   -r   1   -1   postProcessing/residuals/0/
residuals.dat
```

The residuals monitoring can be shown in Fig. 7.18.

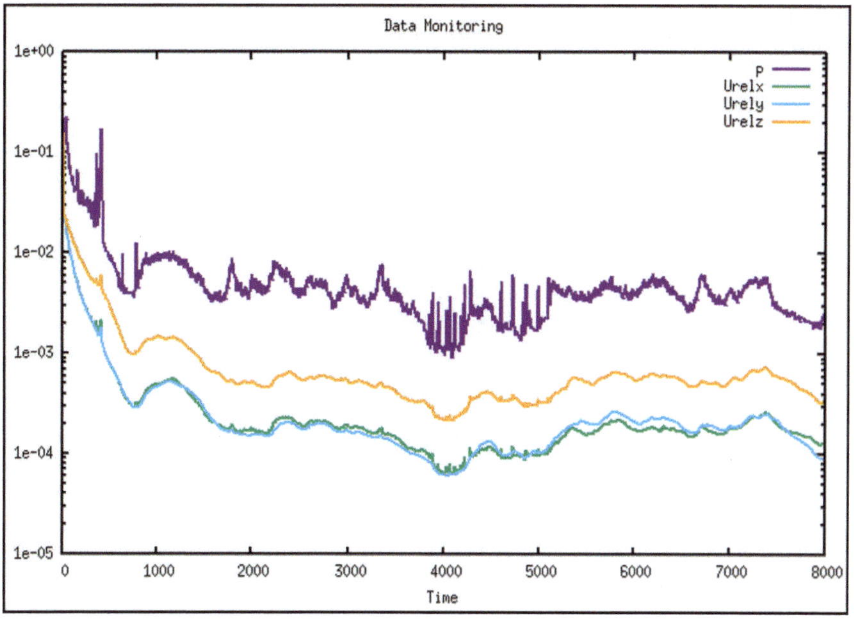

Fig. 7.18 Residuals monitoring

7.1.2 *Results and Post-processing*

7.1.2.1 Data Reconstruct

If parallel calculation is adopted, the data need to be reconstruction, and type reconstructPar in the terminal.

7.1.2.2 Data Read

Type paraFoam to start ParaView, and select all parameters needed as Properties → Fields, as shown in Fig. 7.19.

Press the right button in ParaView as shown in Fig. 7.20 to see the final results.

7.1.2.3 Pressure Contour

Follow the Steps

- Filters → Alphabetical → Extract Block.
- Choose ExtractBlock1.

Fig. 7.19 Parameters selected

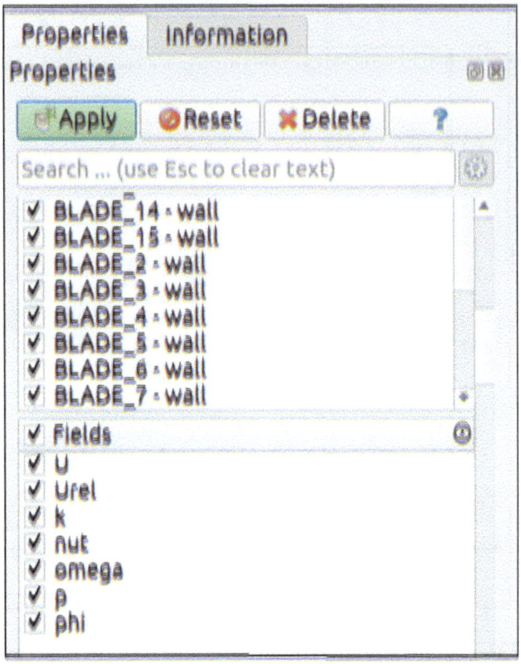

Fig. 7.20 Time chosen

- Properties → Block Indices, choose all blades as shown in Fig. 7.21, and click Apply.
- Properties → Coloring, choose *p*, change the range of *p* into 0–300, and click Apply.
- The pressure contour can be seen in ParaView as Fig. 7.22. One can adjust by rotating, moving, and scaling the model to get a suitable view.

7.1.2.4 3D Streamline

Follow the Steps

- In Pipeline Browser, select runner-clearwater.
- Filters → Alphabetical → Extract Block.
- Press ExtractBlock2.
- Properties → Block Indices---internalMesh, as shown in Fig. 7.23, and Apply.
- Re-select runner-clearwater.
- Filters → Alphabetical → Extract Block.

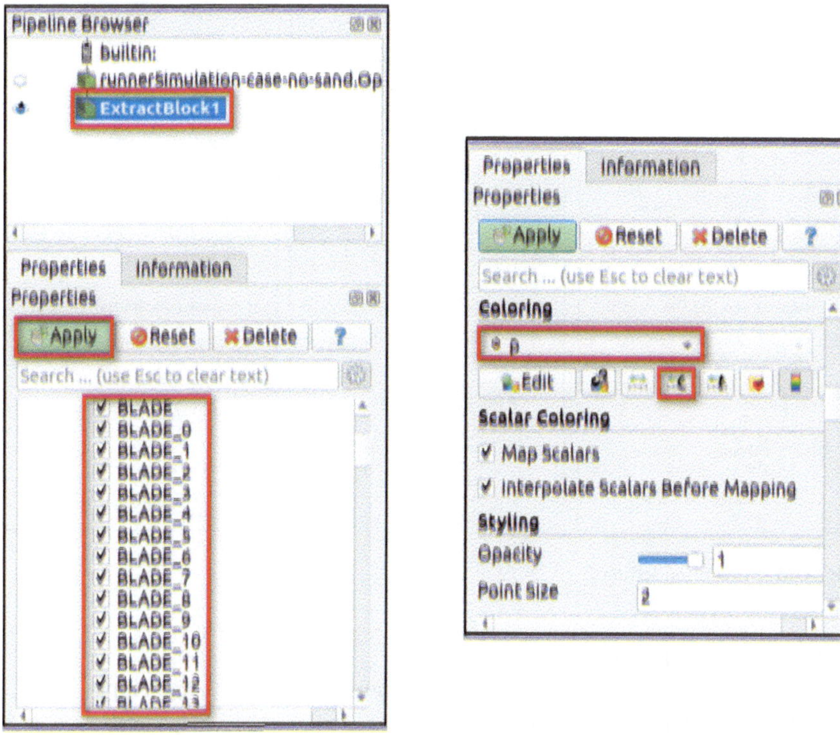

Fig. 7.21 Contour setting

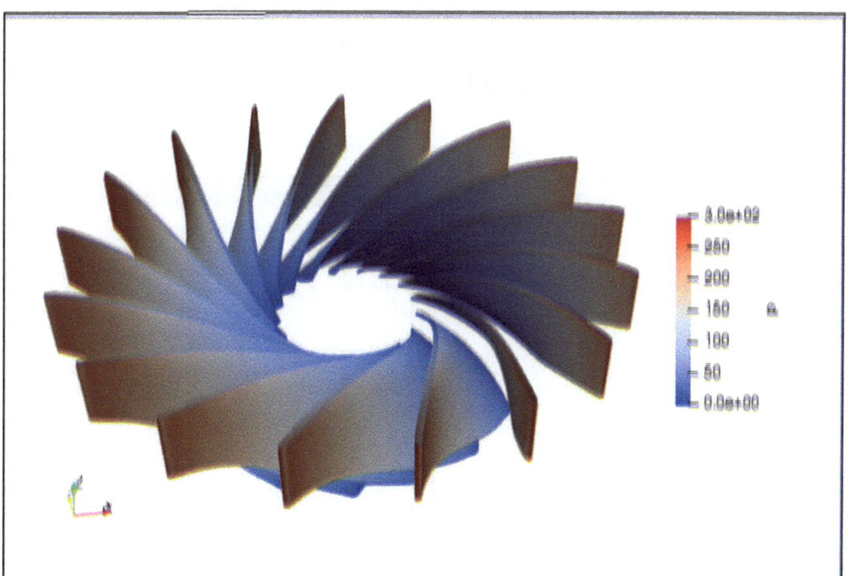

Fig. 7.22 Pressure contour

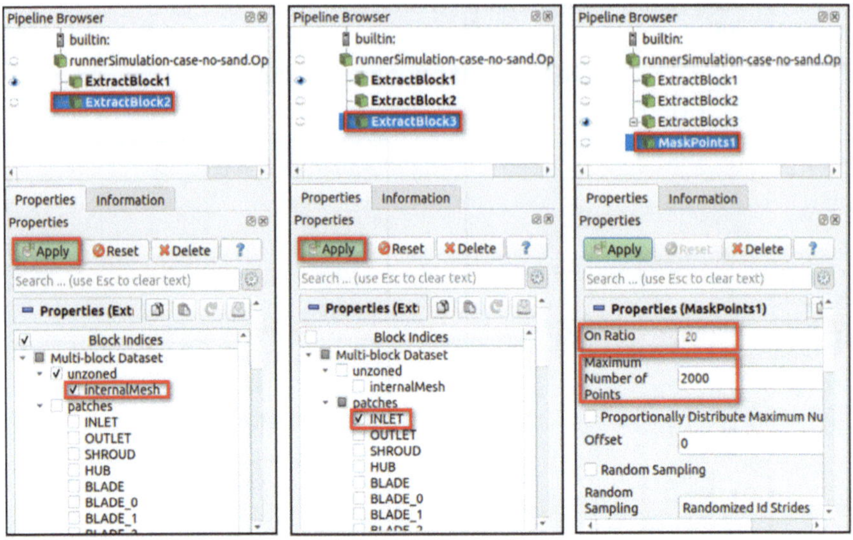

Fig. 7.23 Streamline setting 1

- Press ExtractBlock3.
- Properties → Block Indices–-INLET, as shown in Fig. 7.23, and Apply.
- Filters → Alphabetical → Mask Points.
- In Properties, On Ratio is 20 and Maximum Number of Points is 2000, as shown in Fig. 7.23.
- Filters → Alphabetical → Stream Tracer With Custom Source.
- Input → ExtractBlock2, and Seed Source → MaskPoints1, as shown in Fig. 7.24.
- Properties (StreamTracerWithCustomSource1)→ Vectors → Urel, and the scope is 0–15.

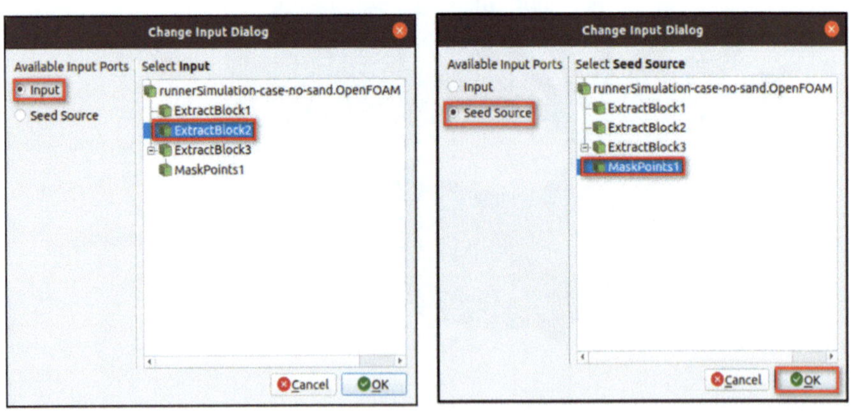

Fig. 7.24 Streamline setting 2

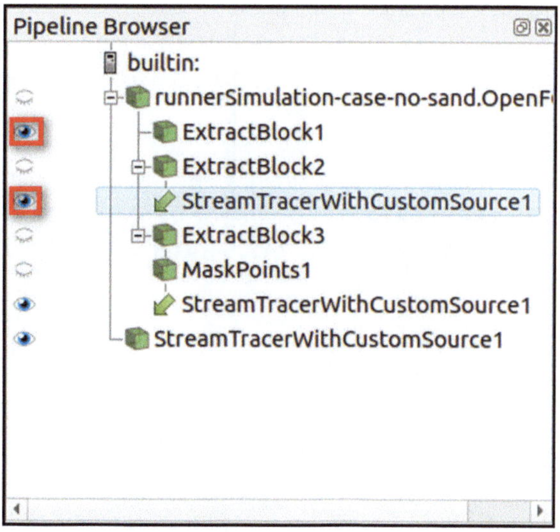

Fig. 7.25 ParaView

- Properties → Styling → Line Width → 2 (open the whole setting by pressing in paraView), select Render Lines As Tubes, and Apply.
- Pipeline Browser → ExtractBlock1 and StreamTracerWithCustomSource1, as shown in Fig. 7.25.
- Both pressure contour and 3D streamline can be viewed now, as shown in Fig. 7.26.

7.2 Erosion Calculation

The present case is especially for the steady state of rotating machines with discrete phase model (DPM) that is applied to OpenFOAM. However, the current version of OpenFOAM has no similar solver, then a revised solver is needed.

7.2.1 Solver Development

The solver development for the present case is based on the existing solver—SRFSimpleFoam. The new solver is named DPMSRFSimpleFoam.

The steps are given as follows:

(a) Open the directory/OpenFOAM-8/applications/solvers/lagrangian, and create a file named DPMSRFSimpleFoam.
(b) Copy files in the /OpenFOAM-8/applications/solvers/incompressible/simpleFoam/SRFSimpleFoam to the directory of DPMSRFSimpleFoam.

Fig. 7.26 Streamline

(c) Change the SRFSimpleFoam.C to DPMSRFSimpleFoam.C. The complete revision is presented as,

```
#include "fvCFD.H"
#include "singlePhaseTransportModel.H"
#include "kinematicMomentumTransportModel.H"
#include "SRFModel.H"
#include "simpleControl.H"
#include "fvOptions.H"
#include "basicKinematicCollidingCloud.H"
#define basicKinematicTypeCloud basicKinematicColliding-
Cloud
//*****************************************/
/
int main(int argc, char *argv[])
{
   argList::addOption
   (
      "cloudName",
      "name",
      "specify alternative cloud name. default is 'kinemat-
icCloud'"
   );
   #include "postProcess.H"
```

```
    #include "setRootCaseLists.H"
    #include "createTime.H"
    #include "createMesh.H"
    #include "createControl.H"
    #include "createFields.H"
    #include "initContinuityErrs.H"
    turbulence->validate();
    //* * * * * * * * * * * * * * * * * * * * * * * * * * * * * * * * * * * /
/
    Info<< "\nStarting time loop\n" << endl;
    while (simple.loop(runTime))
    {
      Info<< "Time = " << runTime.timeName() << nl << endl;
     Info<< "Evolving " << kinematicCloud.name() << endl;
     kinematicCloud.evolve();
     fvVectorMatrix cloudSU(kinematicCloud.SU(Urel));
      // === Pressure-velocity SIMPLE corrector
      {
        #include "UrelEqn.H"
        #include "pEqn.H"
      }
      U = Urel + SRF->U();
      laminarTransport.correct();
      turbulence->correct();
      runTime.write();
     Info<< "ExecutionTime = " << runTime.elapsedCpuTime() << " s"
                                       <<   "     ClockTime    =
" << runTime.elapsedClockTime() << " s"
          << nl << endl;
    }
    Info<< "End\n" << endl;
    return 0;
}
```

(d) Open and edit createFields.H, and change the codes as

```
#include "readGravitationalAcceleration.H"
Info<<"Read transportProperties"<<endl;
IOdictionary transportProperties
(
    IOobject
    (
      "transportProperties",
      runTime.constant(),
      mesh,
      IOobject::MUST_READ_IF_MODIFIED,
      IOobject::NO_WRITE
    )
```

```
);
dimensionedScalar rhoValue
(
    "rhoValue",
    dimDensity,
    transportProperties.lookup("rho")
);
Info<< "rhoValue is " << rhoValue << endl;
Info<< "Creating rho field " << endl;
volScalarField rho
(
    IOobject
    (
        "rho",
        runTime.timeName(),
        mesh,
        IOobject::NO_READ,
        IOobject::NO_WRITE
    ),
    mesh,
    rhoValue
);
Info<< "Reading field p\n" << endl;
volScalarField p
(
    IOobject
    (
        "p",
        runTime.timeName(),
        mesh,
        IOobject::MUST_READ,
        IOobject::AUTO_WRITE
    ),
    mesh
);
Info<< "Reading field Urel\n" << endl;
volVectorField Urel
(
    IOobject
    (
        "Urel",
        runTime.timeName(),
        mesh,
        IOobject::MUST_READ,
        IOobject::AUTO_WRITE
    ),
    mesh
);
Info<< "Reading/calculating face flux field phi\n" << endl;
surfaceScalarField phi
(
```

```
    IOobject
    (
        "phi",
        runTime.timeName(),
        mesh,
        IOobject::READ_IF_PRESENT,
        IOobject::AUTO_WRITE
    ),
    linearInterpolate(Urel) & mesh.Sf()
);
label pRefCell = 0;
scalar pRefValue = 0.0;
setRefCell(p, simple.dict(), pRefCell, pRefValue);
mesh.setFluxRequired(p.name());
Info<< "Creating SRF model\n" << endl;
autoPtr<SRF::SRFModel> SRF(SRF::SRFModel::New(Urel));
// Construct the absolute velocity
volVectorField U
(
    IOobject
    (
        "U",
        runTime.timeName(),
        mesh,
        IOobject::NO_READ,
        IOobject::AUTO_WRITE
    ),
    Urel + SRF->U()
);
singlePhaseTransportModel laminarTransport(U, phi);
autoPtr<incompressible::momentumTransportModel>    turbu-
lence
(
  incompressible::momentumTransportModel::New(U, phi, lami-
narTransport)
);
Info<< "Creating mu field " << endl;
volScalarField mu
(
    IOobject
    (
        "mu",
        runTime.timeName(),
        mesh,
        IOobject::NO_READ,
        IOobject::NO_WRITE
    ),
    rho*laminarTransport.nu()
);
Info<< "Constructing kinematicCloud" << endl;
word kinematicCloudName("kinematicCloud");
```

```
args.optionReadIfPresent("cloudName",    kinematicCloud-
Name);
Info<< "Constructing kinematicCloud " << kinematicCloud-
Name << endl;
basicKinematicTypeCloud kinematicCloud
(
   kinematicCloudName,
   rho,
   Urel,
   mu,
   g
);
#include "createFvOptions.H"
```

(e) Open and edit UrelEqn.H. The complete revision is presented as:

```
// Relative momentum predictor
tmp<fvVectorMatrix> tUrelEqn
(
   fvm::div(phi, Urel)
 + turbulence->divDevSigma(Urel)
 + SRF->Su()
==
   fvOptions(Urel)+(1.0/rho)*cloudSU
);
fvVectorMatrix& UrelEqn = tUrelEqn.ref();
UrelEqn.relax();
fvOptions.constrain(UrelEqn);
if (simple.momentumPredictor())
{
   solve(UrelEqn == -fvc::grad(p));
   fvOptions.correct(Urel);
}
```

(f) Open and add the codes shown below in Make → options:

```
EXE_INC = \
    -I$(LIB_SRC)/lagrangian/basic/lnInclude \
    -I$(LIB_SRC)/lagrangian/intermediate/lnInclude \
    -I$(LIB_SRC)/thermophysicalModels/specie/lnInclude \
```

```
    -I$(LIB_SRC)/thermophysicalModels/basic/lnInclude \
        -I$(LIB_SRC)/thermophysicalModels/reactionThermo/
lnInclude \
    -I$(LIB_SRC)/radiationModels/lnInclude \
    -I$(LIB_SRC)/transportModels/lnInclude \
                    -I$(LIB_SRC)/MomentumTransportModels/
momentumTransportModels/lnInclude \
    -I$(LIB_SRC)/MomentumTransportModels/incompressible/
lnInclude \
    -I$(LIB_SRC)/regionModels/regionModel/lnInclude \
            -I$(LIB_SRC)/regionModels/surfaceFilmModels/
lnInclude \
    -I$(LIB_SRC)/finiteVolume/lnInclude \
    -I$(LIB_SRC)/dynamicFvMesh/lnInclude \
    -I$(LIB_SRC)/meshTools/lnInclude
EXE_LIBS = \
    -llagrangian \
    -llagrangianIntermediate \
    -llagrangianTurbulence \
    -lmomentumTransportModels \
    -lincompressibleMomentumTransportModels \
    -lincompressibleTransportModels \
    -lfiniteVolume \
    -ldynamicFvMesh \
    -ltopoChangerFvMesh \
    -lfvOptions \
    -lmeshTools
```

(g) The final codes are shown as:

```
DPMSRFSimpleFoam.C
EXE = $(FOAM_APPBIN)/DPMSRFSimpleFoam
```

Compile DPMSRFSimpleFoam by typing wmake in the terminal. If the solver has been successfully compiled, a message as shown in Fig. 7.27 will be displayed.

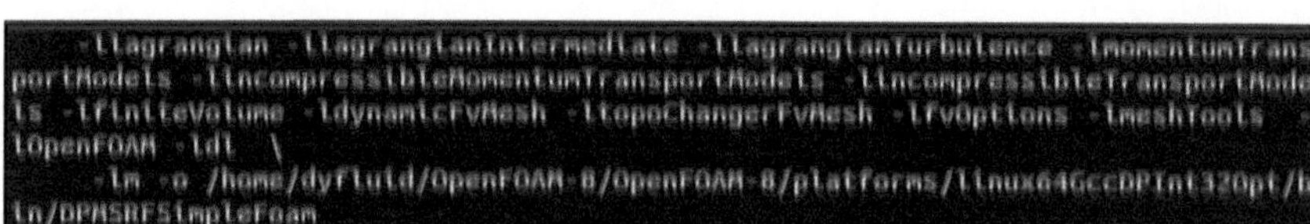

Fig. 7.27 Solver compile

```
transportModel  Newtonian;

nu              [0 2 -1 0 0 0 0] 1e-06;

rho             1000;
```

Fig. 7.28 TransportProperties

```
dimensions      [0 1 -2 0 0 0 0];
value           (0 0 -9.81);
```

Fig. 7.29 File g

7.2.2 Pre-processing

The final data of Sect. 7.1 are taken as the initial state of the present calculation. The steps are as shown,

1. File creation.
2. Physical properties.
3. DPM.
4. Solution and algorithm control.

7.2.2.1 File Creation

Create a file named runner-Erosion, and copy files from runner-clearwater.

7.2.2.2 Physical Properties

Open transportProperties in the directory "/runner-Erosion/constant/", and the complete setting is shown as (Fig. 7.28),

Add g in the directory "constant", and the gravity is changed as (the negative signs means the gravity acts in the negative z direction) (Fig. 7.29).

7.2.2.3 DPM

Add kinematicCloudProperties in the directory "constant", and the complete code is shown as,

```
solution
{
```

```
        active                   yes;
        calcFrequency            20;
        maxTrackTime             5;
        maxCo                    0.4;
        coupled                  true;
        cellValueSourceCorrection off;
        interpolationSchemes
        {
            Urel                 cellPoint;
            rho                  cell;
            mu                   cell;
            DUcDt                cell;
            curlUcDt             cell;
        }
        integrationSchemes
        {
            U                    Euler;
        }
        sourceTerms
        {
          resetOnStartup false;
             schemes
             {
                 U explicit 0.5;
             }
        }
    }
    constantProperties
    {
        rho0 2650;
    }
    subModels
    {
             particleForces
             {
                 sphereDrag;
                 gravity;
                 pressureGradient
                 {
                   U   Urel;
                 }
                 virtualMass
                 {
                   Cvm    0.5;
                 }
                 SaffmanMeiLiftForce
                 {
                    U    Urel;
                 }
                 nonInertialFrame
                 {
```

```
                angularVelocity      (0 0 -104.72);
                centreOfRotation     (0 0 0);
            }
    }
    injectionModels
    {
         model1
         {
             type                       patchInjection;
             parcelBasisType           fixed;
             patchName                INLET;
              U0                       (0 0 0);
             nParticle                  1700; //0.118kg/s
             duration                   1;
             parcelsPerSecond          50000;
             flowRateProfile            constant 1;
             massFlowRate              0;
             sizeDistribution
             {
                 type           fixedValue;
                 fixedValueDistribution
                 {
                      value   0.0001;//m
                 }
             }
         }
    }
    dispersionModel stochasticDispersionRAS;
    patchInteractionModel localInteraction;
    localInteractionCoeffs
    {
         patches
         (
"(SHROUD|HUB|BLADE|BLADE_0|BLADE_1|BLADE_2|BLADE_
3|BLADE_4|BLADE_5|BLADE_6|BLADE_7|BLADE_8|BLADE_
9|BLADE_10|BLADE_11|BLADE_12|BLADE_13|BLADE_14|BLADE_
15)"
             {
                      type       rebound;
                      e          0.97;
                      mu         0.09;
             }
             "(OUTLET|INLET)"
             {
             type       escape;
             }
         );
    }
    heatTransferModel none;
    surfaceFilmModel none;
    collisionModel none;
```

```
        stochasticCollisionModel none;
        radiation off;
    }
    cloudFunctions
    {
    Erosion1
        {
            typeparticleErosion;
             patches(SHROUD HUB BLADE BLADE_0 BLADE_1 BLADE_
    2 BLADE_3 BLADE_4 BLADE_5 BLADE_6 BLADE_7 BLADE_8 BLADE_
    9 BLADE_10 BLADE_11 BLADE_12 BLADE_13 BLADE_14 BLADE_15);
            p   2.7e9;
        }
    Track
        {
            typeparticleTracks;
            trackInterval    3;
            maxSamples    1000000;
            resetOnWrite    yes;
        }
    patchPost
        {
            typepatchPostProcessing;
            maxStoredParcels50000;
             patches(SHROUD HUB BLADE BLADE_0 BLADE_1 BLADE_
    2 BLADE_3 BLADE_4 BLADE_5 BLADE_6 BLADE_7 BLADE_8 BLADE_
    9 BLADE_10 BLADE_11 BLADE_12 BLADE_13 BLADE_14 BLADE_15);
            resetOnWriteyes;
        }
    }
```

The main changes are listed as:

- Solution. It contains setting for particles, such as couplingbetween fluid and particles, interpolationSchemes, and integrationSchemes for numerical scheme of particle equation.
- constantProperties. The density of particle (rho0) is listed here.
- subModels. The models are adopted for calculation, such as particleForces, injectionModels, dispersionModel, and patchInteractionModel. PS: the current setting is just for reference. Suitable values should be input based on real situations.
- CloudFunctions. The models are adopted for post-processing, such as particleErosion, particleTrack, and patchPostProcessing. PS: particleErosion is the model for calculating erosion.

7.2.2.4 Solution and Algorithm Control

- Application

```
application       DPMSRFSimpleFoam;

startFrom         latestTime;

stopAt            endTime;

endTime           12000;

deltaT            1;

writeControl      timeStep;

writeInterval     1000;

purgeWrite        0;

writeFormat       ascii;

writePrecision    6;

writeCompression off;

timeFormat        general;

timePrecision     6;

runTimeModifiable true;
```

Fig. 7.30 ControlDict

The application in controlDict is set as DPMSRFSimpleFoam. The startFrom is set as latestTime, where the data are from the data of clear water. The complete setting is shown as in Fig. 7.30.

- Parallel

  Type decomposePar in the terminal.

- Solution

  Type the following codes for running.

```
mpirun -np 8 DPMSRFSimpleFoam -parallel
```

7.2.3 *Results and Post-processing*

7.2.3.1 Data Reconstruct

Type reconstructPar in the terminal.

7.2.3.2 Data Read

Please ref. to Sect. 7.1.2.

7.2.3.3 Erosion Contour

- Color Map

The way to gain the erosion contour is similar to the pressure contour in Sect. 7.1.2, where the color parameter is kinematicCloud:Q, and the value arrangement is set from 0 to 1e-15.

Color map can be adjusted (Properties → Coloring → Edit → Rainbow Uniform) as shown in Fig. 7.31. The corresponding erosion contour is presented in Fig. 7.32.

- Output

The image can be saved by the ParaView → File → Save Screenshot command as shown in Fig. 7.33.

7.2.3.4 Particle Trajectory

- File of particle trajectory

Copy the file of particleTrackDict under the directory of constant, and the complete setting is shown in Fig. 7.34.

Type steadyParticleTracks in the terminal. A new file named VTK will be created after the operation. The data of particle trajectory is stored in the file.

- Data read by ParaView

ParaView → File → Open, and choose file in the directory /runner-Erosion /VTK/ 12000/particleTracks.vtk.

- Display of particle trajectory

ParaView → Pipeline Browser → particleTracks.vtk.

ParaView → Filters → Alphabetical → Threshold.

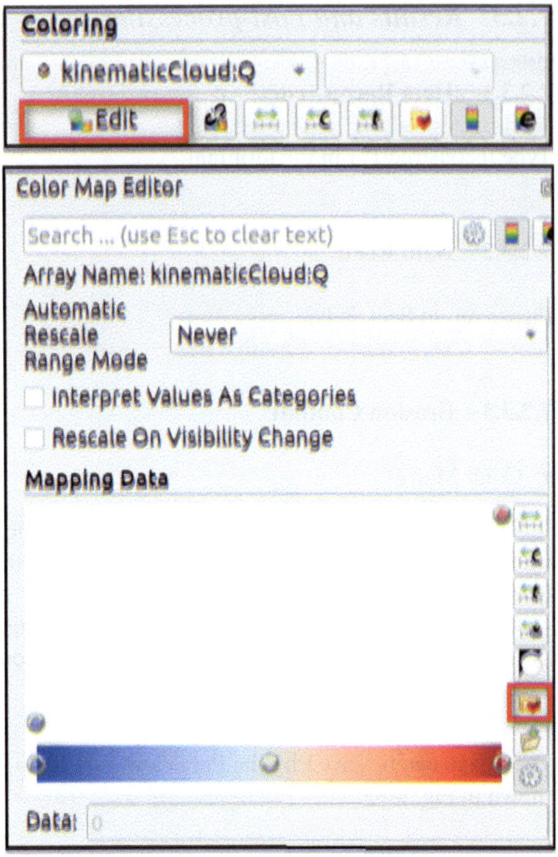

Fig. 7.31 Color map

Properties → Scalars → origId, set the Minimum and Maximum as 0 and 50, which means displaying the particles with the index from 0 to 50, as shown in Fig. 7.35.

Fig. 7.32 Erosion of blades

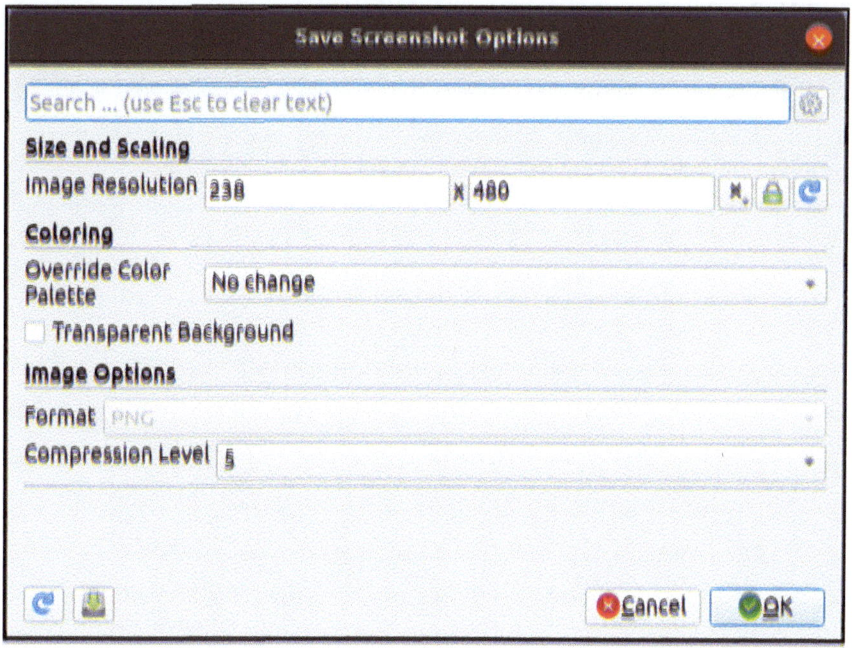

Fig. 7.33 Output of image

```
cloudName        kinematicCloudTracks;

fields ( U origId );
```

Fig. 7.34 ParticleTrackDict

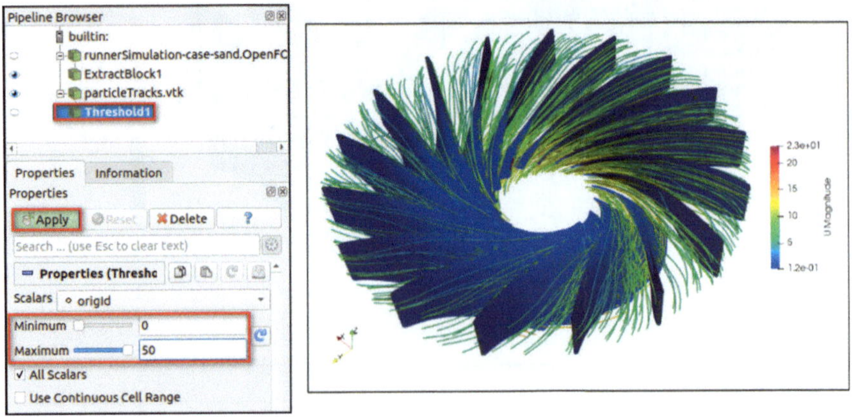

Fig. 7.35 Particle trajectory

Appendices

Appendix A

UEqn.H

```
MRF.correctBoundaryVelocity(U); //BC correction in a moving frame
of reference
        tmp<fvVectorMatrix> tUEqn
        (
            fvm::div(phi, U) //convective part
          + MRF.DDt(U) //Coriolis forces in a moving frame of reference
          + turbulence->divDevSigma(U)* //stress part, incompressible flow,
including both molecular viscosity and turbulent (or eddy) viscosity,
         ==
             fvOptions(U) //additional forces on the RHS of UEqn
        );
        fvVectorMatrix& UEqn = tUEqn.ref();
```


Z. Guo et al., *Design of Francis Turbine and OpenFOAM Methods*,
https://doi.org/10.1007/978-981-99-8381-0

UEqn.relax(); //under relax the momentum equations in order to prevent the solution to diverge

fvOptions.constrain(UEqn); //constrain U under required values

if (simple.momentumPredictor())

{

solve(UEqn == -fvc::grad(p)); //solve the system to get the estimate $\mathbf{U}_{\mathbf{P}}^{*}$ after adding the contribution of the pressure gradient to the right hand side of the matrix

fvOptions.correct(U); //corrections of velocity

}

* Refer to Appendix B.

pEqn.H

{

volScalarField rAU(1.0/UEqn.A()); //rAU; $\mathrm{UEqn.A()} = A_p$

volVectorField HbyA(constrainHbyA(rAU*UEqn.H(), U, p)); //HbyA, $\mathrm{UEqn.H()} = \sum A_N U_N$

surfaceScalarField phiHbyA("phiHbyA", fvc::flux(HbyA)); //phiHbyA, surface flux of HbyA

MRF.makeRelative(phiHbyA); //make the given absolute velocity relative within the MRF region.

adjustPhi(phiHbyA, U, p); //ensure zero net flux in the entire area. If inflow disagrees with outflow, adjust the assignable speed boundary, in order to make the entire system flux conservation

tmp<volScalarField> rAtU(rAU); //Scalar field rAtU

if (simple.consistent()) //SIMPLEC

{

rAtU = 1.0/(1.0/rAU - UEqn.H1());//$1/(Ap + \sum A_N U_N)$

phiHbyA +=

fvc::interpolate(rAtU() - rAU)*fvc::snGrad(p)*mesh.magSf();//$HbyA_{p,SIMPLEC,s}$

HbyA -= (rAU - rAtU())*fvc::grad(p); //$HbyA_{p,SIMPLEC}$

}

tUEqn.clear(); //clear tUEqn

```
        // Update the pressure BC (fixedFluxPressure boundary
condition) to ensure flux consistency
        constrainPressure(p, U, phiHbyA, rAtU(), MRF);

        // Non-orthogonal pressure corrector loop
        while (simple.correctNonOrthogonal())
        {
            fvScalarMatrix pEqn
            (
                fvm::laplacian(rAtU(), p) == fvc::div(phiHbyA)
//pressure equation for the SIMPLEC algorithm
            );

            pEqn.setReference(pRefCell, pRefValue); //set reference
point and value

            pEqn.solve();

            if (simple.finalNonOrthogonalIter()) //Flag to indicate the
last non-orthogonal iteration
            {
                phi = phiHbyA - pEqn.flux(); //correction of flux
            }
        }
```

```
#include "continuityErrs.H"

// Explicitly relax pressure for momentum corrector

p.relax();


// Momentum corrector

U = HbyA - rAtU()*fvc::grad(p);

U.correctBoundaryConditions(); //BCs corrector

fvOptions.correct(U); //Velocities corrector when additional forces are considered

}
```

Appendix B

turbulence- > divDevSigma(U).

The general deviatoric shear stress term provided by the MomentumTransport-Models library is named divDevTau for compressible flow and divDevSigma (sigma = tau/rho) for incompressible flow, and the spherical part of the shear stress is assumed to be either included in the pressure or handled separately. The corresponding stress function sigma is also provided which in the case of Reynolds stress closure returns the effective Reynolds stress (including the laminar contribution) or for other Reynolds averaged or large eddy turbulence closures returns the modeled Reynolds stress or subgrid stress respectively.

For Reynolds Averaged Navier–Stokes equations, the Reynolds stress is given under Bousinessq Hypothesis,

$$-\rho\overline{u_i'u_j'} = 2\mu_t S_{ij} - \tfrac{2}{3}\mu_t \tfrac{\partial U_k}{\partial x_k}\delta_{ij} - \tfrac{2}{3}\rho k\delta_{ij}$$

For incompressible flows,

$$-\rho\overline{u_i'u_j'} = 2\mu_t S_{ij} - \frac{2}{3}\rho k\delta_{ij}\left(\text{where } \frac{\partial U_k}{\partial x_k} = 0\right)$$

u_i' is the fluctuating components, k is the turbulent kinetic energy, defined by

$$k = \frac{1}{2}\overline{u_i'u_i'}$$

and S_{ij} is a symmetric tensor, defined as,

$$S_{ij} = \frac{1}{2}\left(\frac{\partial U_i}{\partial x_j} + \frac{\partial U_j}{\partial x_i}\right)$$

Then,

$$-\rho\overline{u_i'u_j'} = \mu_t\left(\frac{\partial U_i}{\partial x_j} + \frac{\partial U_j}{\partial x_i}\right) - \frac{2}{3}\mu_t\frac{\partial U_k}{\partial x_k}\delta_{ij} - \frac{2}{3}\rho k\delta_{ij}$$

$$-\rho\overline{u_i'u_j'} = \mu_t\frac{\partial U_i}{\partial x_j} + \left(\mu_t\frac{\partial U_j}{\partial x_i} - \frac{2}{3}\mu_t\frac{\partial U_k}{\partial x_k}\delta_{ij}\right) - \frac{2}{3}\rho k\delta_{ij}$$

The vector form is

$$-\rho\overline{u'u'} = \mu_t\nabla \boldsymbol{U} + \left(\mu_t(\nabla \boldsymbol{U})^T - \frac{2}{3}\mu_t(\nabla\cdot \boldsymbol{U})\boldsymbol{I}\right) - \frac{2}{3}\rho k\boldsymbol{I}$$

For incompressible flows, the term $\frac{2}{3}\rho k\boldsymbol{I}$ is usually combined with the pressure gradient term by defining a turbulent pressure p_{tur} as

$$p_{tur} = p_l + \frac{2}{3}\rho k$$

Then

$$p = \frac{p_{tur}}{\rho}$$

The terms of $\mu_t \nabla \boldsymbol{U} + \left(\mu_t (\nabla \boldsymbol{U})^T - \frac{2}{3}\mu_t (\nabla \cdot \boldsymbol{U})\boldsymbol{I}\right)$ can be combined with that of the mean flow as

$$(\mu_t + \mu)\left[\nabla \boldsymbol{U} + \left((\nabla \boldsymbol{U})^T - \frac{2}{3}(\nabla \cdot \boldsymbol{U})\boldsymbol{I}\right)\right]$$

Let

$$\frac{(\mu_t + \mu)\left[\nabla \boldsymbol{U} + \left((\nabla \boldsymbol{U})^T - \frac{2}{3}(\nabla \cdot \boldsymbol{U})\boldsymbol{I}\right)\right]}{\rho} = \nu\left[\nabla \boldsymbol{U} + \left((\nabla \boldsymbol{U})^T - \frac{2}{3}(\nabla \cdot \boldsymbol{U})\boldsymbol{I}\right)\right]$$

$$\nu = (\mu_t + \mu)/\rho$$

The turbulence- > divDevSigma(U) is

$$\nabla \cdot \left\{\nu\left[\nabla \boldsymbol{U} + \left((\nabla \boldsymbol{U})^T - \frac{2}{3}(\nabla \cdot \boldsymbol{U})\boldsymbol{I}\right)\right]\right\}$$

Notice that for incompressible flows $\nabla \cdot \boldsymbol{U} = 0$ and $\nabla \cdot \left((\nabla \boldsymbol{U})^T\right) = 0$.
Then,

$$\nabla \cdot \left\{\nu\left[\nabla \boldsymbol{U} + \left((\nabla \boldsymbol{U})^T - \frac{2}{3}(\nabla \cdot \boldsymbol{U})\boldsymbol{I}\right)\right]\right\} = \nabla \cdot (\nu \nabla \mathrm{U})$$

Appendix C

alphaEqn.H

```
// Set the time blending factor, 1 for Euler
scalar cnCoeff = 1.0/(1.0 + ocCoeff);

// Standard face-flux compression coefficient
surfaceScalarField
phic(mixture.cAlpha()*mag(alphaPhic/mesh.magSf())); //cAlpha is
defined in the <case>/system/fvSolution file

// Add the optional isotropic compression contribution
if (icAlpha > 0) //icAlpha is defined in the <case>/system/fvSolution
file

    {
        phic *= (1.0 - icAlpha);
        phic += (mixture.cAlpha()*icAlpha)*fvc::interpolate(mag(U));
    }

// Add the optional shear compression contribution
if (scAlpha > 0)
    {
        phic +=
```

```
            scAlpha*mag(mesh.delta() &
    fvc::interpolate(symm(fvc::grad(U))));
        }

      surfaceScalarField::Boundary& phicBf = phic.boundaryFieldRef();

      // Do not compress interface at non-coupled boundary faces
      forAll(phic.boundaryField(), patchi)
      {
            fvsPatchScalarField& phicp = phicBf[patchi];

            if (!phicp.coupled())
              {
                phicp == 0;
              }
          }

      tmp<surfaceScalarField> phiCN(alphaPhic);

      // Calculate the Crank-Nicolson off-centred volumetric flux
      if (ocCoeff > 0)
          {
              phiCN = cnCoeff*alphaPhic + (1.0 -
```

```
cnCoeff)*alphaPhic.oldTime();
        }

        if (MULESCorr) //MULES algorithm
        {
            #include "alphaSuSp.H"
            fvScalarMatrix alpha1Eqn//alpha1Eqn
            (
                (
                    LTS
                    ?
fv::localEulerDdtScheme<scalar>(mesh).fvmDdt(alphac, alpha1) //local
time-step
                    : fv::EulerDdtScheme<scalar>(mesh).fvmDdt(alpha1)
//Euler
                )
              + fv::gaussConvectionScheme<scalar>
                (
                    mesh,
                    phiCN,
                    upwind<scalar>(mesh, phiCN)
                ).fvmDiv(phiCN, alpha1) //convective term
              - fvm::Sp(fvc::ddt(alphac) + fvc::div(phiCN), alpha1)
```

```
        ==
            Su + fvm::Sp(Sp + divU, alpha1)
    );

    alpha1Eqn.solve(); //solve alpha1Eqn

    Info<< "Phase-1 volume fraction = "
        << alpha1.weightedAverage(mesh.Vsc()).value()
        << "   Min(" << alpha1.name() << ") = " <<
min(alpha1).value()
        << "   Max(" << alpha1.name() << ") = " <<
max(alpha1).value()
        << endl;

    tmp<surfaceScalarField> talphaPhi1UD(alpha1Eqn.flux());
    alphaPhi10 = talphaPhi1UD();

    if (alphaApplyPrevCorr && talphaPhi1Corr0.valid())
     {
        Info<< "Applying the previous iteration compression flux"
<< endl;
          MULES::correct
              (
```

```
                    alphac,
                    alpha1,
                    alphaPhi10,
                    talphaPhi1Corr0.ref(),
                    zeroField(), zeroField(),
                    oneField(),
                    zeroField()
                );

            alphaPhi10 += talphaPhi1Corr0();
        }

    // Cache the upwind-flux
     talphaPhi1Corr0 = talphaPhi1UD;

alpha2 = 1.0 - alpha1;

mixture.correct();
    }
```

UEqn.H

```
MRF.correctBoundaryVelocity(U);

fvVectorMatrix UEqn
(
    fvm::ddt(alphacRho, U) //temporal term
  + MRF.DDt(alphacRho, U) //Coriolis forces in a moving frame of
reference
  - fvm::Sp(fvc::ddt(rho) + fvc::div(rhoPhi), U) //process for a
better boundedness
  + fvm::div(rhoPhi, U) //convection term
  + turbulence->divDevRhoReff(U) //stress part, incompressible
flow, including both molecular viscosity and turbulent (or eddy) viscosity
 ==
    fvOptions(rho, U) //additional forces on the RHS of momentum
equations
  + cloudSU //source term, momentum transfer with particles
);

UEqn.relax(); //under relax the momentum equations in order to
prevent the solution to diverge
fvOptions.constrain(UEqn); //constrain U under required values
```

```
    volScalarField rAUc(1.0/UEqn.A());
    surfaceScalarField rAUcf(fvc::interpolate(rAUc)); //rAUcf
obtained from rAUc interpolation

    surfaceScalarField phicForces
    (
        (fvc::interpolate(rAUc*cloudVolSUSu) & mesh.Sf()) //explicit
part of cloudSU
    );


    if (pimple.momentumPredictor())
    {
        solve
        (
            UEqn
         ==
            fvc::reconstruct
            (
                phicForces/rAUcf
              +
                (
                        fvc::interpolate
```

```
                    (
                         mixture.sigmaK()
                    )*fvc::snGrad(alpha1) //surface tension term
                    - ghf*fvc::snGrad(rho) //volumetric force term
                    - fvc::snGrad(p_rgh) //pressure term (modified
pressure by rearrangement)
                ) * mesh.magSf()
            )
    );
```

//solve the system to get the estimate $\mathbf{U}_{\mathbf{P}}^{*}$ after adding the contribution of the pressure gradient to the right hand side of the matrix

```
        fvOptions.correct(U); //corrections of velocity
    }
```

pEqn.H

```
{
        volVectorField HbyA(constrainHbyA(rAUc*UEqn.H(), U,
p_rgh));
        surfaceScalarField phiHbyA
        (
            "phiHbyA",
            fvc::flux(HbyA)
          + alphacf*fvc::interpolate(rho*rAUc)*fvc::ddtCorr(U, phi)
//Choi Correction
        );

        MRF.makeRelative(phiHbyA); //make the given absolute
velocity relative within the MRF region.
        adjustPhi(phiHbyA, U, p_rgh); //ensure zero net flux in the
entire area to make the entire system flux conservation

        surfaceScalarField phig
        (
            phicForces +
            (
fvc::interpolate(mixture.sigmaK())*fvc::snGrad(alpha1) //surface tension
```

```
term
                    - ghf*fvc::snGrad(rho) //volumetric force term
                )*rAUcf*mesh.magSf()
            );

            phiHbyA += phig; //update phiHbyA

                constrainPressure(p_rgh, U, phiHbyA, rAUcf, MRF);
//Update the pressure BCs to ensure flux consistency

            while (pimple.correctNonOrthogonal()) //Non-orthogonal
corrector loop.
            {
                surfaceScalarField Dp("Dp", alphacf*rAUcf);

                fvScalarMatrix p_rghEqn
                (
                    fvm::laplacian(Dp, p_rgh)
                  ==
                    fvc::ddt(alphac) + fvc::div(alphacf*phiHbyA)
//pressure equation
                );
               p_rghEqn.setReference(pRefCell, getRefCellValue(p_rgh,
```

```
pRefCell)); //set reference point and value

p_rghEqn.solve(mesh.solver(p_rgh.select(pimple.finalInnerIter())));
//solve pressure equation

            if (pimple.finalNonOrthogonalIter())
            {
                phi = phiHbyA - p_rghEqn.flux()/alphacf;
//correction of flux

                p_rgh.relax(); //under relax the pressure equation

                U =
                    HbyA
                  + rAUc*fvc::reconstruct((phig -
p_rghEqn.flux()/alphacf)/rAUcf); //Momentum corrector

                U.correctBoundaryConditions(); // BCs corrector

                fvOptions.correct(U); //Velocities corrector when
                additional forces are considered
```

```
            }
        }

        #include "continuityErrs.H"

        p == p_rgh + rho*gh;

        if (p_rgh.needReference())
        {
            p += dimensionedScalar
            (
                "p",
                p.dimensions(),
                pRefValue - getRefCellValue(p, pRefCell)
            );
            p_rgh = p - rho*gh;
        }
    }
```